THE BLACK BOOK OF BOOBYTRAPS

BY

LYLE WHITNEY

Desert Publications

El Dorado, AR 71730 U. S. A

Black Book of Boobytraps

by

Lyle Whitney

Published by Desert Publications
215 S. Washington Ave.
El Dorado, AR 71730 U.S.A.
1-870-862-3811
info@deltapress.com

Printed in U.S.A
ISBN: 978-0-87947-149-1

10 9 8 7 6 5 4 3 2

Desert Publication is a division of
The DELTA GROUP, Ltd.

Direct all inquiries & orders to the above address.

WARNING NOTICE

Neither the author nor the publisher assumes any responsibility for the use or misuse of the information contained in this book. The author and publisher specifically disclaim any personal liability, loss, or risk incurred as a consequence of the use and application, either directly or indirectly, of any advice or information presented herein.

Technical data presented herein reflects the author's individual beliefs and experience with particular equipment and environments under specific circumstances which the reader can not duplicate or experience exactly. The information in this book should therefore be used for guidance only and should be approached with great caution.

Since neither the author nor the publisher have control over the building materials used, use of techniques or equipment, or the abilities of the reader, no responsibility, either implied or expressed, is assumed for the use or misuse of the data or procedures given in this book.

Table of Contents

INTRODUCTION

Imagine that you are standing by a window. Outside. Looking through a gap between lace curtains. Dim lights inside make everything show up like a three-dimensional movie screen. Your eyes are on a woman that has just put on a night gown. You are confident that you are invisible to her, standing as you are in the darkness, As you always have been. She knows little or nothing about you, but you know much about her. You have studied her carefully. You have seen the man pack luggage in a car. You have witnessed the parting ritual. You know she is alone in the house. You feel an almost overwhelming drive to smash through and attack her this instant. But maintain control, knowing it will be better to be patient, and less direct.

The light goes off, and you stand waiting. Images of you kneeling over her come to mind. Her throat is in your hands, and you are slowly choking the life out of her as she begs for her life. You feel embodied by the lustful energy that always precedes and accompanies a kill. And decide to make this one last longer than the others. You decide that you must have her tonight.

On the darkest side of the house, a screen has stupidly

been left off. You open the window easily. Push the curtain inward, and look down, turning on a flashlight just long enough to see that you have a good surface to land on. A weak-looking table with glass bottles on it will have to be avoided, but that will be no problem. You still have a clean shot to the floor. There is even a throw rug, to make a quiet landing easier.

You crawl through, feet first, on your belly. Descend slowly, body pressed flat against the wall. And touch down softly. As you put weight on your feet, you are slightly startled as something seems to give way, as though there had been a cardboard box or something under the rug. For an instant it occurs to you to wonder what might have been under the rug. Then suddenly a knife blade rips into your back and nails you to the wall.

You die slow, unable to move, wondering who or what killed you.

Section I

Straight Deadfalls

CHAPTER 1

Boobytrapping Theory & General Information

The Nature of the Warrior

A boobytrap is a warrior that jumps suddenly out of nowhere and attacks viciously. Like a religious fanatic, it is suicidal: it will end its existance in one violent act. It is fearless and invulnerable, and so virtually unstoppable. It has no feelings and knows nothing, and so cannot be tortured or questioned.

As a guard it will kill, wound, slow, or warn away an enemy passing through a defended area. Even if it misses the target, it will at least slow his progress. Only the most hardened and determined individual will move quickly through an area known to be boobytrapped. That individual will be hit by the second or third guard.

As an alarm set at access points it can serve to warn of the approach of an enemy. In groups it can guard a perimeter. It can startle or injure an attacker, and so make him more vulnerable to counter-attack.

As an assassin it is set on the enemy's territory. It will make one or more kills, then die. In groups it can harass an enemy, destroy his organization, and lower his strength. One or more can immediately precede or accompany an ambush.

Hitting a Target

Inability to think is the boobytrap's primary weakness. The decision to attack is made by action of the target, and the attack is of a pre-set form and direction. Aim and timing cannot be re-set to fit the target. All must be calculated, judged, and set beforehand.

Correctness of setting depends largely on how much is known about the target. Position, shape, speed, and direction of travel must be known or judged. The mechanic must empathize with his intended target.

Paths and confining passageways are the most common places for traps to be set, because target position can easily be judged. Either can put a target in line with a payload. Often they are the best places to set a trap. But for this reason they are also the places the target will be most suspicious of. If the target does not suspect a trap, doors and windows are ideal settings. But if a trap is suspected to be at a door or window, there is a good chance it will be discovered. It can quickly and easily be rendered harmless by a heavy object thrown onto the trigger with the target safely out of the way. This can be dealt with by covering an area other than or in addition to what the trap appears to cover.

Obstacles (i.e. log on highway, overturned table) and objects of interest (souvenier, weapon, bound or dead comrade) can be used to control target position and align it with a payload. A trap can be baited by attaching a trigger line to anything a target might want. Making an area relatively easy to travel through (i.e. one door left unlocked) will increase the chance of the target entering that area.

Traps can be used in combination, with one controlling target position and the other delivering the kill. An obvious false trap can direct a target toward a hidden, real one. A pit can stop and hold a target as a payload falls. A loud noise (i.e. shot, door slamming) or rushing sound can make a person dive for cover, and the best cover area from that point can be boobytrapped.

Concealment

Any part seen by the target can render a trap useless. The importance of concealment and how extensive it must be depends largely on how fast the target will be traveling, and whether or not it suspects a trap. A person who kicks in a door and jumps inside shooting will not see a trap in time to avoid it. Someone going slowly and carefully and looking for traps will only be hit by one that is very well hidden and designed. Someone moving at average speed who does not suspect a trap will probably be hit by one that is moderately concealed.

Anything done by the mechanic to make a target unaware of a trap is a form of concealment. It can be arrived at in two ways: invisibility and disguise.

The easiest—and usually best way—to make a trap invisible is to set all or part of it outside the target's field of vision. A trap set on one side of a door or window will be invisible from the other side. Objects can be stood or hung between the target and trap.

Rigging lines can be turned and made to run flat against existing structures such as walls or limbs. They can be placed behind drywall and paneling, along studs and joists, through walls, and above ceilings. They can also be hidden through or behind foliage. Things like curtains and tapestries can be hung over them. Smallness and being the same color and texture as the background are good aids to concealment.

Things can be set between the target and payload providing they do not interfere with payload descent. This can be foliage, a false ceiling, insulation, and things of that nature. Payloads can hide behind lamp fixtures, ceiling fans, rafters, shelves, hanging clothes or towels. A swinging payload can travel around things used to hide it.

Insubstantial objects such as curtains and unlatched doors can be pushed out of the way or penetrated by a sufficiently powerful payload driven by good rigging. A trigger can be set so that if the concealing object is disturbed, the trap fires.

A tripline must be set in the target's path. It can be hidden by something like tall grass, but usually it must be set in the target's field of vision. In such cases being both very thin and the same color and texture as the background are essential to invisibility.

Coloring can be done with paint, ink, dye, or food coloring. The aim of good camouflage is to break up the pattern and make it blend with the background. How this is done depends on the nature of the background.

If the mechanic is skilled enough to set a hair trigger, sewing thread makes an excellent tripline. It is very thin and comes in a variety of colors. Very thin, camouflaged steel wire is manufactured for use by the military as triplines. Fishing line is usually designed to be very visible, and so must usually be colored to make it invisible.

A pressure plate trigger can be hidden very easily. Something like a throw rug or dirty laundry can be laid over the top of it. It can be set in a shallow pit and covered with dirt or leaves. One or more loose floorboards can make a pressure plate.

Total darkness will of course make a trap invisible. Near darkness will make one less visible. Shadows in otherwise lighted areas can hide trap components. If it can be known that a target will be moving through darkness (i.e. an enemy making a night raid), extensive concealment is unnecessary. A person moving through darkness with a flashlight will either move very slowly, probably miss a moderately concealed trap, and be hit by it. The driver of a vehicle traveling at night sees only what the headlights illuminate. An area can be guarded to a certain extent by the knowledge that an intruder must either have light (in which case you can see him), or be stumbling in darkness through a boobytrapped area. During daylight the same traps are visible, and so relatively safe.

Traps are disguised by being made to look like something else. If the target does not suspect a trap, the disguise can look slightly out of place and still work, so long as the trap does not look like an obviuos trap. For example, a target might wonder why an electrical cord is running across a ceiling at a weird angle, but not suspect it of being boobytrap rigging.

Triggers can be disguised as things that the target will want or have to move. This can be a chair blocking a doorway, or a log laying across a road. Anything a target might want to pick up (money, weapon), or manipulate in any way (light string, curtain, strong box, door) can be a disguised trigger. Any such object should look natural, as though there is a reason for it being where it is. For example, a rifle laying at the bottom of a flight of stairs can be set to look as though it had fallen down the stairs.

Payloads are difficult to disguise and must usually be hidden. One may be made to look like a light fixture, stereo speaker, something sitting on an overhead shelf, and so on. Payloads hanging in trees can be made to look like bird nests.

One way to make certain that disguised trap parts, or objects used to hide trap parts do not look out of place is to make everything in the area look out of place. The more disordered an enviroment is, the less able a target will be to make any kind of sense of it. If the disorder is so extensive and so creative as to convince the target that it was done by an insane person, so much the better. That will increase any psychological effect the traps will have.

If a house is a total mess, with laundry laying all over, a few articles positioned so as to hide trap components will not look out of place. If several strange-looking objects are hanging from a ceiling, the thought that one of them might be a disguised payload is less likely to occur than if one is hanging alone over a strategic point. If furniture is turned over and scattered haphazardly, a chair blocking a doorway can look perfectly natural. Curtains, blankets, and similar articles hanging from walls and ceilings and over doorways can make a house as easy to hide traps in as a jungle. If many lines run back and forth across a room or doorway, any of them might be a tripline or rigging. Anything hanging or sitting overhead might drop straight or swing down in any direction. Anything standing might tip over. Anything laying on the floor might be covering a pressure plate. Anything might be a trigger.

Bombed out areas and places where large battles have taken place are naturally disordered. Construction sites and many industrial areas tend to be very disordered. Places like jungles, forests, and farmyards seem disordered to city people. These areas are prime settings for boobytraps.

In some cases it will be necessary to conceal the fact that a trap ever existed, in other words, make a boobytrap look like an accident. In the case of firetraps this can be done by using only flammable materials for trap components, and insuring that the fire is hot enough and of enough duration to consume them. Sometimes traps can be made to look like accidents caused by stupidity on the part of the target. The mechanic must be able to look at and understand his enviroment well enough to know how accidents might happen, and then cause them to happen.

The last aspect of concealment is good housekeeping. Scraps, marks, foot prints, and anything else that can indicate that some sort of construction took place must be disposed of.

Deadfall Forms

Two boobytrap mechanisms derive power initially from gravity: deadfalls and pitfalls. Deadfalls are those that drop a payload onto the target. Though the mechanism can sit anywhere in relation to the ultimate target, it will always contain something that will fall downward toward an objective.

All deadfall mechanisms have three principal components: a payload, a trigger, and rigging. These are its body, mind, and energy, respectively. As with a human warrior, the parts must be well balanced and coordinated.

There are three primary deadfall forms: a freefall, a pendulum, and tilt. They will be referred to in the text as straight, swinging, and standing, respectively (fig. 1). In each case a payload is held suspended until released, at which time it falls.

Each form has many manifestations, all within the general framework and with certain things in common. Once the basic structure and certain principles are understood clearly, the devices that each form can be adapted to are unlimited.

Tools and Construction Techniques

A boobytrapper is a construction mechanic. Methods and so-called "tricks of the trade" used by construction tradesmen apply equally well to boobytrapping. Studying books on construction and learning from experienced tradesmen will do much to improve boobytrapping ability. Skills that apply well to boobytrapping include: engineering, carpentry, rigging, electrical wiring, woodcraft and other primitive skills.

Basic construction tools such as hammers, welders, hoists, and so on will all be used in boobytrapping. But a boobytrap mechanic will often be limited to tools that can easily or inconspicuously be carried. These tools should be chosen carefully. They should be of good quality and design. Versatility is a key asset.

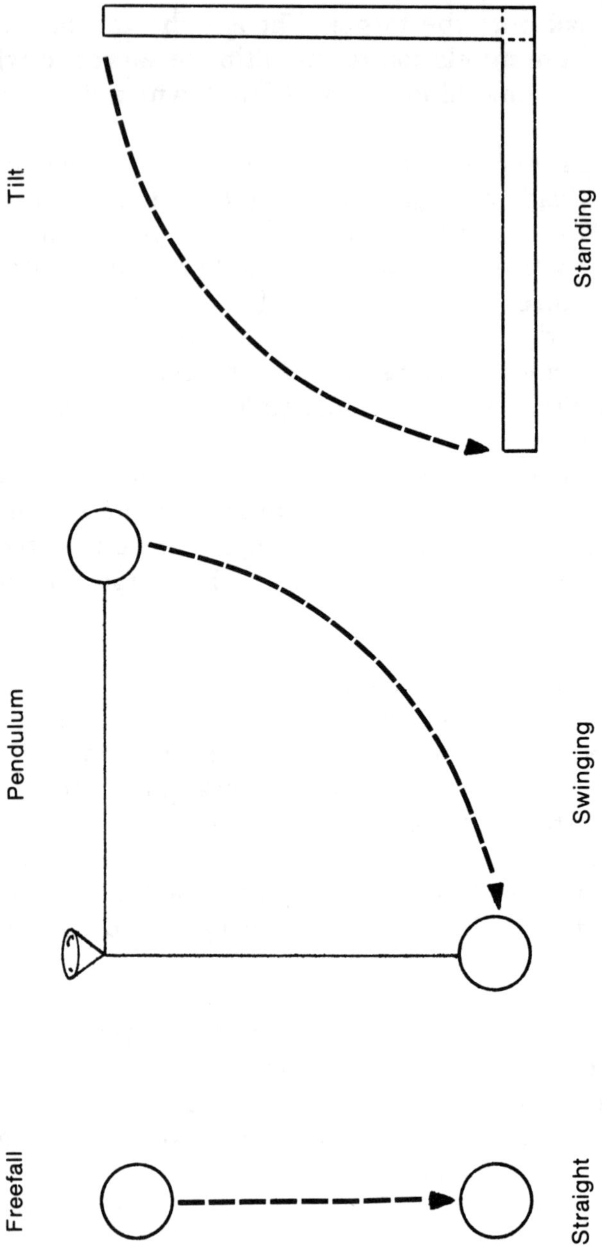

Fig. 1. The three deadfall forms.

Pliers and Vice Grips

A pair of vice grips is one of the most useful and versatile of all boobytrap tools. It can be used to clamp trap components such as trigger parts down so that precise work can be done on them quickly and easily. It can cut and strip wire, act as a wrench, and pull nails. With something heavy clamped in its jaws (i.e. large bolt) it can be a hammer. With something sharp in its jaws (i.e. nail) it can punch or gouge holes in walls and ceilings. And it can do anything pliers can do.

It can clamp and hold a load line securely to a pulley while a trigger is being set, which makes it very useful to a mechanic working alone. It can also function as various trap components, either temporarily or permanently. This includes anchors, pulleys, and triggers.

Pliers will have substantially more power than a pair of vice grips of the same size, and are easier to work with. Pliers are less versatile than vice grips, but no less useful. Other forms of pliers include bolt cutters, channel locks, and side cutters. These can all be very useful to a boobytrapper.

Hammers

A hammer is probably the oldest tool known to man, and it is still one of the most useful, especially in construction. A boobytrap mechanic will often require a hammer of some sort. Anything hard and heavy, such as rocks, bricks, crescent wrenches, or cooking pots, can be a hammer.

Obviously the best hammer is one designed to be a hammer. For most boobytrap construction, a carpenter's claw hammer is by far the best. It is a combination hammer and wrecking bat. Carpenters use it more than any other tool, because it is so powerful in regard to building or taking apart anything made out of wood. The hammer makes installation of hardware (pulleys, anchors, stakes, etc.) very fast and easy; the claw can produce material by pulling nails, or ripping boards loose.

Most claw hammers are not designed for heavy prying, and can be broken if too much pressure is put on them; those with steel handles are unlikely to break, and so are preferable to all others.

A single-edged ax is a hammer on one side. A hand ax is to a woodsman what a claw hammer is to a carpenter. It is very powerful in regard to shaping (cutting, splitting, notching) wood, and is especially useful in wilderness areas. Some come with notches for pulling nails, which can be very useful.

Saws

Next in importance to a hammer for wood construction is a saw. Folding tree saws designed for camping are easy to carry and can rip through wood very quickly. Small flexible saws with finger rings for handles, also designed for camping, are very easy to carry, and can shape trigger parts and cut notches quickly and easily. Knives with saw backs (pilot survival knife) can be similarly useful. Some swiss army knives have saw blades, which work surprisingly well.

Knives

Knives are used for cutting cordage, sharpening and otherwise shaping wood. The best ones are those designed to be tools rather than weapons. A pilot's survival knife is the best for most purposes. It is well designed to give a person power to shape his enviroment. Its functions as a saw and hammer can be very useful. Swiss army knives, especially the ones with saw blades, can be very useful.

Pry Bars

A pry bar is one of the most useful and versatile of all

construction tools. It is not uncommon for knife blades to be broken because of being used as pry bars.

A flat-ended screwdriver makes a very good pry bar, and can also be used to punch holes in thin walls and ceilings (for use as pulleys, to hide rigging, etc.), in addition to removing or installing screws. A plier's handle can be ground or filed (preferably filed) to a screwdriver-like end. This can function more or less like a screwdriver, and makes a very good pry bar.

Plumb Bob

A tool used to draw a straight line between two vertical points is called a plumb bob. This can be very useful in aligning payloads with targeted areas. Any object that can hang at the end of a line and hold it tight can be a plumb bob. The payload itself makes an excellent plumb bob, but is not always practical to use. The more important precision is, the more important it is that the end be small and centered directly under the line. A double-edged dagger with a hole in the handle hanging at the end of a string makes as good a plumb bob as anything money can buy. A rock will give an approximate reading.

The tool is used by holding the string at pulley height, and letting the weight (bob) float slightly above the targeted area. The string draws a straight line between the point targeted and the position the pulley needs to be.

In some cases it will be easier to raise and drop a weight, and mark where it hits.

Tool Belt

It is common for construction hands to carry tools and material on a belt. This is especially true of workers who have to climb to do their work, such as carpenters and ironworkers. A mechanic rigging deadfalls will often have to climb to do his work.

The belts are designed to give easy access to tools

without inhibiting movement. They vary from trade to trade, but most have a few things in common. Long tools hang on either side, and one or two bags hang on the back (for nails, bolts, etc.). The front is left empty to allow the legs to bend forward. Such a belt can be very useful to a boobytrapper.

A military web belt with a butt pack makes a very good tool belt. The belt can carry a hammer or ax, knife, bolt cutters, canteen (which can carry water, liquid fuel, or anything else), and other necessary tools. Ammunition or other pouches designed to be attached to the belt can carry smaller tools and materials. Grenade carriers found on many ammunition pouches make excellent hammer carriers.

The pack can carry a large quantity of rigging lines, a block and tackle, shackles, rope slings, tie wire, glue, tape, and igniters.

The belt and pack are designed to be comfortable and allow freedom of movement. They are very similar to the tool belts with bolt bags worn by ironworkers when climbing and working on structural steel. If tools are arranged correctly they will not get in the way when climbing. Suspenders are a useful improvement.

Lashing

Many trap components are shaped by being lashed together. Spikes are lashed to weights, arms are hinged with lashing, matches are lashed around strikers, igniters are lashed to fuel containers, and so on. Most lashing can be done quickly and easily with tape. The better quality the tape is, the easier and faster it will be to work with, and the better and stronger the lashing will be done. The best kinds of tape for lashing are electrical, duct, packaging (the stranded kind), and medical.

Lashing can also be done with string, cordage, or any rigging material. Strips cut from clothing, blankets, furniture, etc., make very good lashing material. If saturated with glue they can have a great deal of strength and holding power, and be used more or less like tape. Glue by itself makes good lashing.

Primitive people used lashing extensively. Animal skin turned into rawhide is very strong, and if applied wet shrinks as it dries, which causes it to hold very securely. Sinew (strands of tendons) is very strong and pliable. Hair braided into cordage is not exceptionally strong, but is suitable for lashing. Small young trees (one-eigth to one-quarter inch diameter) can be twisted until the fibers separate, then cut loose from the ground and used for lashing. For more information on primitive lashing, consult books on primitive and wilderness survival skills.

The most common lashing material on construction sites is wire. The small diameter wire usually called "tie wire" makes lashing very fast and easy. Wire is the strongest of all lashing materials.

Temporary Rigging

Setting a trigger can be difficult or impossible if the load line has tension on it. It can also be dangerous, because the mechanic must often occupy the targeted area while setting the trigger. If the payload is held in place by something other than the load line, setting the trigger will be easy and safe.

If two or more mechanics are working together, one or more can support the load while the other sets the trigger.

Vice grips can support a load by clamping a loadline to a pulley or some other object. The line should not be pulled away from the position it will be in when the trap is set, or the load will drop some distance after the clamp is removed. The clamp should not be set so tight as to cut or damage the line. A softener (piece of leather, cloth) should be placed between the line and jaw.

A sling or rope can be attached to the payload and temporarily secured to the pulley. It should lead from the same approximate point on the payload that the permanent load line will lead from. It then holds the load as the trigger is set.

Temporary rigging can remain in place as a safety. Removing it arms the trap.

Scaffolds and Ladders

Since deadfalls will usually be set overhead, a ladder or scaffold will often be necessary. This can usually be made from materials in the immediate enviroment. Furniture can be moved and stood on. Doors can usually be removed easily by taking out hinge pins. A scaffold can then be made by laying the door between supports such as chairs, desks, shelves, or window frames. It should be covered with something like a blanket or rug to keep it from being scratched. It can then be re-installed, and look perfectly innocent.

Logs can be leaned against tree trunks or limbs and used as ladders. Two or more can be laid between limbs to form a scaffold. A reasonably good ladder can be made by cutting notches in a log, or by lashing sticks on perpendicular to it. Any ladder can be improved by being tied-off, either at the top or the bottom.

Safety

When the subject of boobytrapping is considered, the first thing that usually comes to mind is the possibility of the trapper falling into his own trap. This is a very real possibility. It can happen as a result of forgetting about a trap that has been set, or as an accident during construction.

Any construction is dangerous, and the nature of boobytrapping makes it among the most dangerous of all. Since for the most part boobytraps are temporary one-shot devices, and because smallness of rigging materials can be a key factor in concealment, lines and anchors are used to near limits of their strength. This practice is universally recommended against in safety manuals when the lines will be supporting any kind of load and where safety is a factor. The danger is compounded in boobytrapping, because the loads are designed to attack.

It is a common safety practice on construction sites to stay out from under any hanging load, no matter how secure it seems to be. Hanging boobytrap loads (payloads) are relatively insecure, since it is their designed purpose to fall. Since triggers are set in the approximate area the payload is aimed to attack, the mechanic will be in or near the targeted area (under the load) when setting the trigger. This is the most likely time for the payload to fall accidently.

This danger can be lessened a great deal by securing the load to an anchor with something independent of the load line and trigger assembly while the trigger is being set. That acts as a positive safety, which can be left on until the mechanic leaves the area. Ways of doing this are discussed under the heading **Temporary Rigging**. Even with such a safety, the mechanic should spend as little time as possible under the load. He should remain concious of the load and the course it will take at all times.

Systems designed to start fires are potentially extremely dangerous to the builder. Some of the best fuels are very volatile (mix with air readily, becoming highly flammable and/or explosive), and can be ignited prematurely in a number of ways. Volatile fuels held in an open container can fill a room with fumes and turn it into a bomb, given the right conditions. Some systems involve filling open-topped containers with liquid fuel and lighting the top of the fuel, with the container set to spill. If the air has enough fumes in it, that could be fatal.

There will sometimes be a question of when to fill a

vessel with fuel. A charged fuel and igniter system is potentially very dangerous to handle, especially when doing something like climbing. Filling a vessel after the trigger has been set can be equally dangerous. The decision of when to charge the system must be made according to the nature of the trap and the enviroment it is set in, the skill of the mechanic, time available, and other factors. Attaching the pull cord on the igniter to its anchor should be the last thing done.

Poisons, acids, firearms, heavy weights, and spikes can be as dangerous to handle as fuels and igniters. A small slip on the part of the mechanic can result in a serious accident.

The above considerations are all in addition to typical dangers of conventional construction, such as falling, being hit on the head by a dropped tool, and many others.

As with all construction, the most important factor of safety is alertness on the part of the mechanic. Skill and understanding of the mechanism he is building is near equal in importance. This can be developed by building models before attempting to build real ones.

The mechanisms in this book were designed with little or no consideration given to safety, it being subjugated to power, concealability, and simplicity. They are a safety man's nightmare. Given that, you should understand and remember that your own safety is always in your own hands.

The other aspect of safety is not falling into the trap after construction is complete. This obviously is avoided by remaining aware of the trap, especially the location and nature of the trigger and kill zone. A trap set on friendly territory can be forgotten about with the passing of time. Such a trap should be taken down or put on safe (payload tied to an anchor) whenever possible. Concealment can be moved to make traps conspicuous, and so relatively safe. Traps concealed by darkness will be relatively safe during daylight. Traps can be identified by signs or other markings.

Ethics

The subject of boobytrapping raises broad ethical questions, the most serious of which is that the eye of the builder will not be on the target when the trap is activated. Once set, it attacks indiscriminately; so there is a chance that the trap will attack an innocent person. For this reason many people consider boobytraps to be necessarily unethical, believing that there is no such thing as a good boobytrap.

Boobytraps are not the only weapons that kill indiscriminately. Land mines lay hidden and wait for a target to activate them just like boobytraps do. Bombs dropped from planes and long range artillery are aimed in the general direction of an enemy, but the eye on the shooter is not on the target. These weapons are all known to arbitrarily kill innocent civilians. In spite of the awesome frequency of reports of things like hospitals and orphanages being bombed, only the most hardcore liberal considers bombing to be unethical in and of itself.

It is not so much the fact that these weapons are used by legitimate governments in legitimate wars that makes them seem ethical in spite of their occasional random killing. It is that their use is supposed to be limited (at least in theory) to situations where not using them would result in a greater evil than the random loss of life they might produce. In such a situation, the use of boobytraps can be equally ethical.

Nevertheless, it is the responsibility of the mechanic to do everything in his power to insure that only a correct target is hit.

A trap set on enemy territory will be likely to hit only a correct target. The more the area is inhabited exclusively by the enemy, the less likely it is that an innocent person will be hit. The office, home, or vehicle of a specific enemy is a better place for a trap than the general area he inhabits. Traps set in an area being evacuated which will soon be overrun by enemy forces will be all but certain to hit only correct targets.

A trap set to kill potential burglars might hit a person who has a legitimate reason to break into an area. For example, if a fire starts inside a house, someone (neighbor, fireman) might break in to put it out. In an emergency situation, someone might break into a house to use a telephone.

Except for the relatively high possibility of hitting a wrong target, boobytraps are the same as all other weapons and tools as far as ethics are concerned.

Legal Implications

As far as the law is concerned, all boobytrapping is criminal. Except in a war zone, there is no such thing as a legal boobytrap. Anyone building a boobytrap with a purpose to injure or kill should be well aware that the law considers him to be a criminal for doing so, and should act accordingly.

CHAPTER 2

Deadfalls

A straight deadfall sits directly over the target area and drops straight. The area it covers is what lies between the payload and ground, at any elevation. This means that the target can be standing, sitting, or crawling, and still be in the kill zone.

Since gravity pulls it straight down, this payload will drive straight into the ground below it. Providing the ground is level, it will drive in at a 90 degree angle. This enhances power considerably, because the ground keeps the target from flowing away with the force. The payload and ground act like the arm and platform of a mousetrap, respectively, with the ground being the surface that the payload smashes the animal against.

In its pure form, the primary target of a straight deadfall is the top of the head and shoulders. This is true of all forms that do not transfer energy to another mechanism.

The mechanism consists of a payload, a load line, a pulley, and a trigger. The pulley and load line hold the payload poised over the kill zone charged with potential energy. When activated, the trigger releases the line, and drops the payload.

CHAPTER 3

TRIGGERING

The trap will be activated by the target manipulating some sort of trigger. The trigger will be kicked, stepped on, pushed, picked-up, or moved in some other way. It will be set in the target's path, as a pressure plate or trip-line, or something he will want to pick up (tool, weapon, souvenier), or something he will want or need to move (obstacle, vehicle).

Since contact between the target and trigger is what causes release, the trigger determines where and to a certain extent in what position the target will be when attacked. In most cases it will be approximately under the payload, depending on what the target will be doing when attacked (crawling, walking, running, kicking in a door, snooping around, driving).

That, along with the nature of the surroundings, the amount of time and available resources, will determine the type of trigger to be used.

No matter what form the mechanism takes, it must have the following qualities: 1) sufficient strength to support the weight of the payload; 2) the ability for the loadline to be attached in such a way that it can be released by the amount of pressure the target will put on it; 3) reasonable certainty that it will not be missed without it covering an area larger than is covered by the payload; 4) reliability. It should be as simple as possible.

The trigger itself, or tripline, can be hard or soft. A hard trigger can act as a lever (treadle), and make a hair trigger easy to set. Many variations of this type can only be activated from one direction, which may or may not be a disadvantage. It can be some kind of stick, like a broom or pool cue, or a platform, like a shelf board. A soft trigger can be made of string, electrical cord, fishing line, colored steel wire made specifically for boobytrapping, sewing thread, cotton or nylon string.

One end is usually attached to a dead anchor. This needs only to be secure enough to support the weight of the trigger, plus what must be applied to effect release. A hard trigger can often be simply laid on top of something, even the ground. In some cases the end can be free floating. A soft trigger must either be attached to something, or be activated by being pulled by the target (i.e. a false light string).

The other end of the tripline holds the load line secured to an anchor, and in most cases is in turn held in place by the force between them. This is due primarily to the fact that good triggers are designed to detach from the load line after triggering has occurred. Depending on how the rigging is done, this trigger will support some percent of the deadfall's weight. It must be fastened securely to something substantial, such as a wall stud, heavy furniture, appliance, tree, stake, or vehicle.

Timing

In many cases rate of fall will need to be calculated, and compared to target travel speed. Disregarding resistance to air (which for our purposes is not a significant factor), and drag (which is not a factor unless rigging is done poorly), all payloads will fall at the same speed, regarless of their weight. A freefalling body accelerates at a constant rate as it falls. A simple formula used to find how far a body will fall in a given amount of time is: $D = 16T^2$ with D and T being the distance in feet and time in seconds, respectively. So in one second (D=16x1x1), a body will fall sixteen feet; in two seconds (D=16x2x2) it will fall sixty-four feet. Usually it will be the distance that is known rather than the time. If the payload will fall ten feet, for example, the formula would read: $10 = 16T^2$. The answer is T=.79, or approximately three quarters of a second. In most cases approximate calculations will be sufficient.

Average walking speed ranges from three to seven feet per second. Crawling speed will be less. Running speed can reach a maximum of around thirty feet per second, but in most cases should be substantially less. The slower travel speed is, the easier and more certain timing is, because the target will spend more time in the kill zone.

Fall time and travel speed must be compared to determine where the trigger sits in relation to the payload. It should be remembered that the foot of a walking person always precedes his head.

Targeting vehicles is slightly more complicated because of their greater range of speed. But if approximate speed can be estimated, they can still be either hit, wrecked, or stopped with a good degree of certainty. Approximate speed can usually be estimated by the nature of the road, type of vehicle, and other factors.

Vehicles on rough roads and off-road areas travel slow, and so can be targeted in about the same way as walking targets. Areas that force a vehicle to move very slowly, such as sharp curves, holes, or deep mud are ideal locations. A covered pit can stop and hold a vehicle in place as a payload falls, and so eliminate timing as a factor altogether.

Vehicles can be wrecked by large and heavy objects dropped in front of them. Obviously the faster they are going the more effect the payload will have. Greater speed also makes timing easier and more certain, because as speed increases minimum stopping distance increases. Bad road conditions, such as rain, snow, and ice, further simplify timing by extending minimum stopping distance. Oil or sand spread on a road will have the same effect.

The payload is set to touch down immediately in front of a vehicle traveling the maximum estimated speed. If a trap is set for a vehicle with a maximum speed of one-hundred miles per hour, a vehicle traveling fifty miles per hour will still be wrecked and one traveling forty will have to stop very suddenly, and might be wrecked. Tolerance of error in estimating speed in this case is plus zero minus fifty miles per hour. As maximum estimated speed decreases, so does minimum speed. Tolerance decreases dramatically with lower speeds, because stopping distance is shorter. If the payload is one that can easily be steered around, tolerance is even less.

To determine the distance between the trigger and payload, payload travel time is compared to vehicle speed, which should be translated into feet per second. One mile per hour equals approximately one and a half (1.4667) feet per second. A vehicle traveling seventy-five feet in one second (fifty miles per hour, or approximately seventy-five feet per second) will travel seventy-five feet in the time it takes a sixteen foot high payload to touch down. So the distance between the payload and trigger for a sixteen foot high payload and maximum vehicle speed of fifty miles per hour should be seventy-five feet.

If vehicle speed cannot be estimated to within a permissible tolerance, two or more payloads can be used to be certain of a hit. Both are triggered simultaneously, and the payloads are placed some distance apart. A slow vehicle will hit or be hit by the first payload; a fast one will miss the first and hit the second.

Timing can be set so as to allow a vehicle to escape from one pursuing it. The first vehicle hits the trigger and passes safely under the falling payload. The payload then touches down in front of the second vehicle.

TRIGGERS

Pin and Eye

This is among the simplest, and often best ways for this type of trap to be triggered. It can be strong, dependable, and very adaptable, because it has a variety of forms. The amount of pressure needed to cause release can be adjusted in a few ways.

The best thing for the eye is a screw eye, as sold in hardware stores. If one is not available, a nail can be bent to form an eye. Two parallel nails can act as an eye, providing the angle of pull is correct. If no hardware of that type is available, an eye can be made by punching, drilling, burning, or cutting a hole or deep notch in a stud, or two holes in something like panelling. A forked stake can be driven into the ground. A creative mechanic will discover that many areas have pre-existing eyes (drawer handles, spaces between boards and bricks, wire mesh fences). In any case, the eye must be large and smooth enough so the line can run freely out of it, and secure enough to support the weight.

The pin must be rigid, reasonably strong, and smooth. A nail works well. Smoothing out the end with a grinder or stone will make it work better. The head will keep the tripline from pulling off the end. Needles and pins work well. A sliver of wood (toothpick) will work if it does not need to support too much weight. It can be strengthened considerably by being saturated with something like superglue. Screw drivers, ice picks, tooth brushes, and many kitchen utensils work well.

A loop is tied in the free end of the load line. It must be one that will not slip, like a bowline. An easy and fast way to tie a knot that will work just as well (for this purpose, not for general rigging) is to double the end by folding a few inches of line back, and tie a half hitch. A better way is to double the line and tie both ends together. This will make a loop that is free of any knot, and so can run freely through small pulleys without catching.

To set the trigger, the loop is barely passed through the eye, and the pin is passed through it, going from one side to the other. As long as the pin is in contact with both sides of the eye, the loop will not pull through. Once it is pulled past one point, it will release. A hair trigger is set by having the pin go through just far enough to barely be under both sides of the eye. (Fig. 2)

If necessary, the trigger action can be smoothed by placing rollers between the pin and one or both sides of the eye. A smooth object (magazine) can be placed between the pin and eye, or between the line and any rough surface. (Fig. 3) Graphite (pencil lead), soap, grease or oil can be rubbed on the pin.

The tripline is attached (glued or tied) to the pin and set so that pressure on it pulls the pin out. It is important that the angle of pull be correct. The pin must be pulled straight out. Otherwise it will bind rather than release. (Fig. 4) The angle of pull can be turned with a pulley.

The load line can come in to the eye from any angle, providing the eye is a complete circle.

It is possible to reduce the amount of force between the pin and the eye, even to the point that the pin will fall out by itself. This is another way to make a hair trigger, and is also useful when only a weak pin such as a toothpick, pencil, or candle is available. The idea is to let the eye take some or most of the weight. This is done by wrapping the load line part way or completely around one side of the eye before passing the pin through it. (Fig. 5) In some ca-

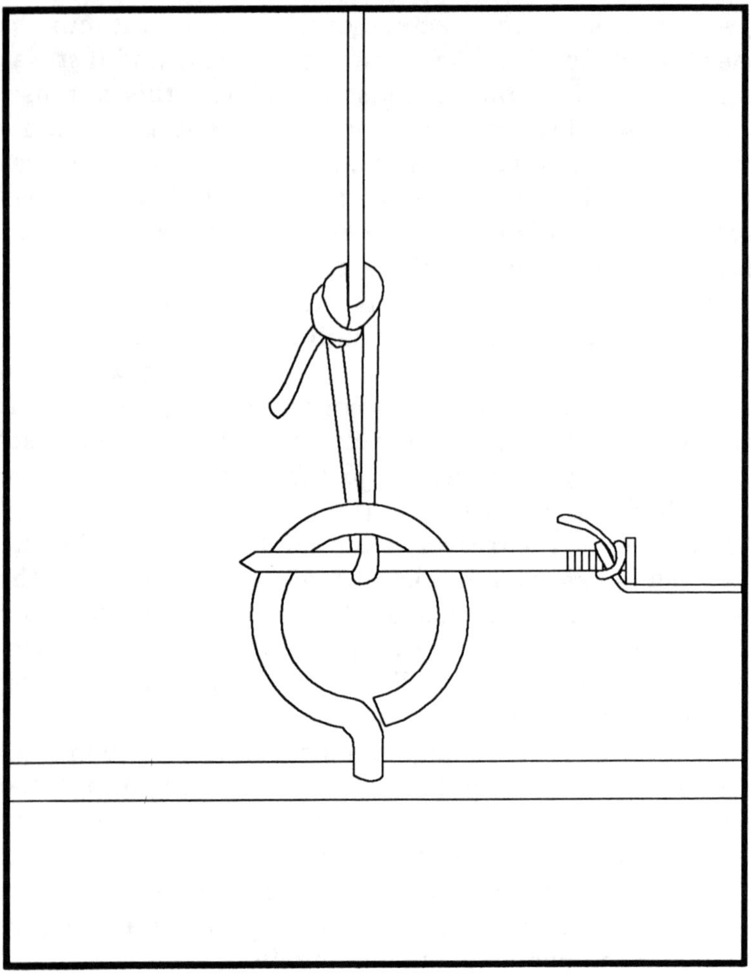

Fig. 2. Pin and Eye with hair trigger.

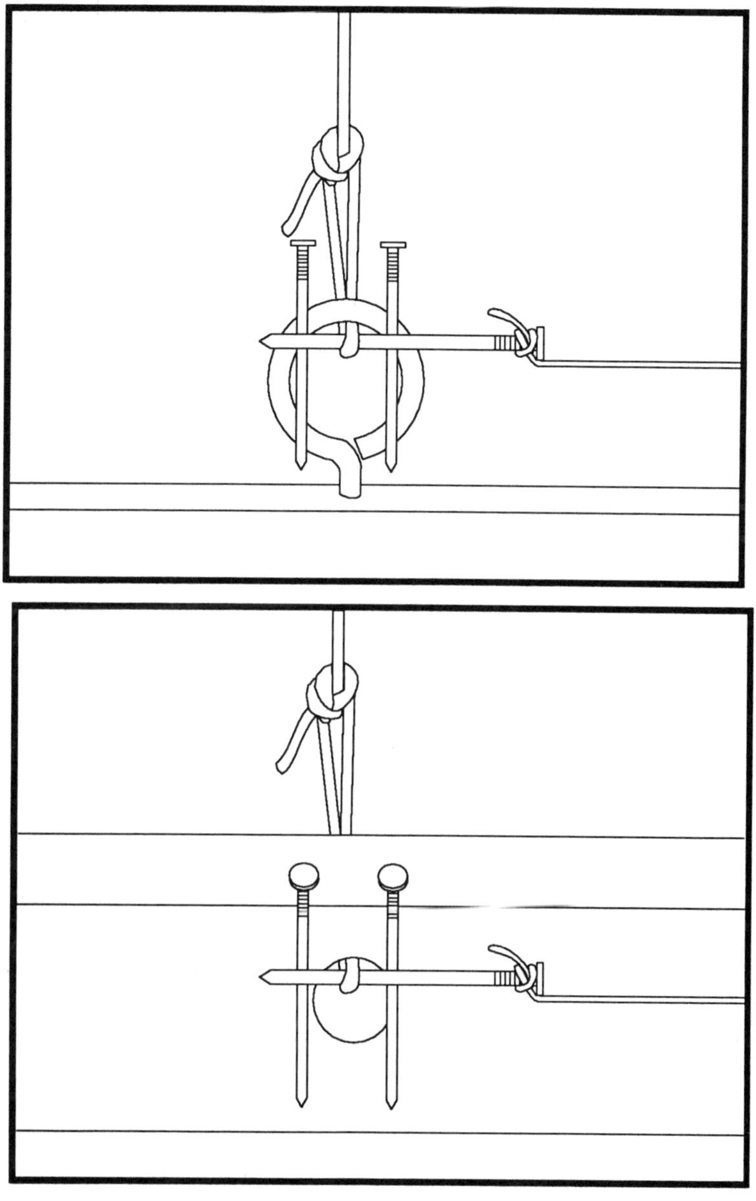

Fig. 3. Rollers can be used to smooth trigger action. Note that this trigger is set more or less on safe.

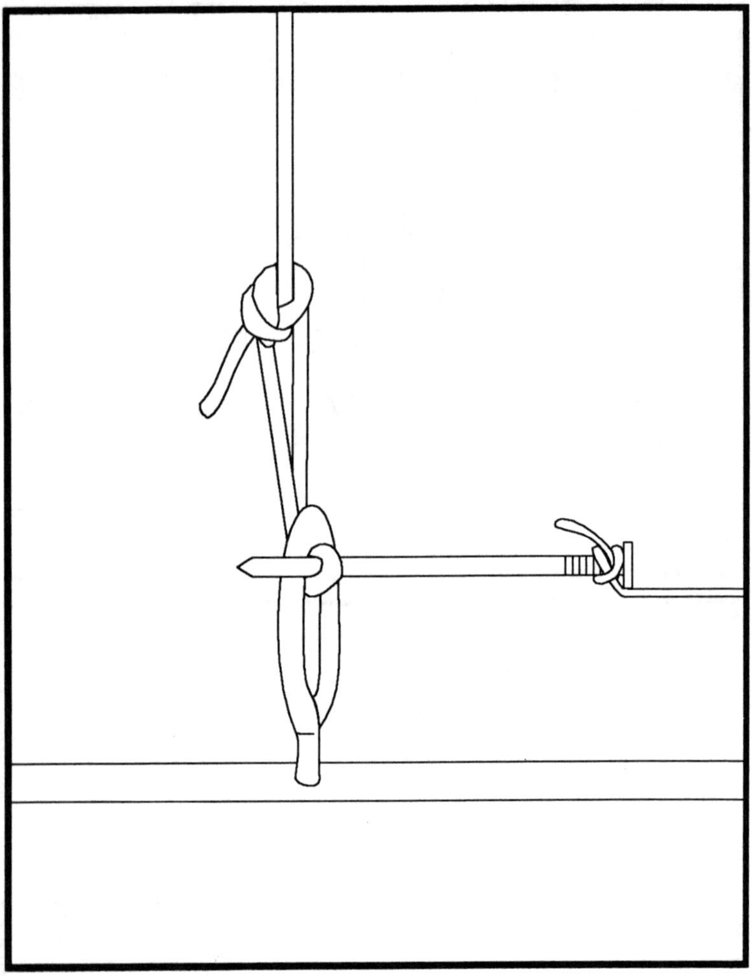

Fig. 4. The pin is binding because it is not being pulled straight out. It will not release unless a great deal of pressure is put on it.

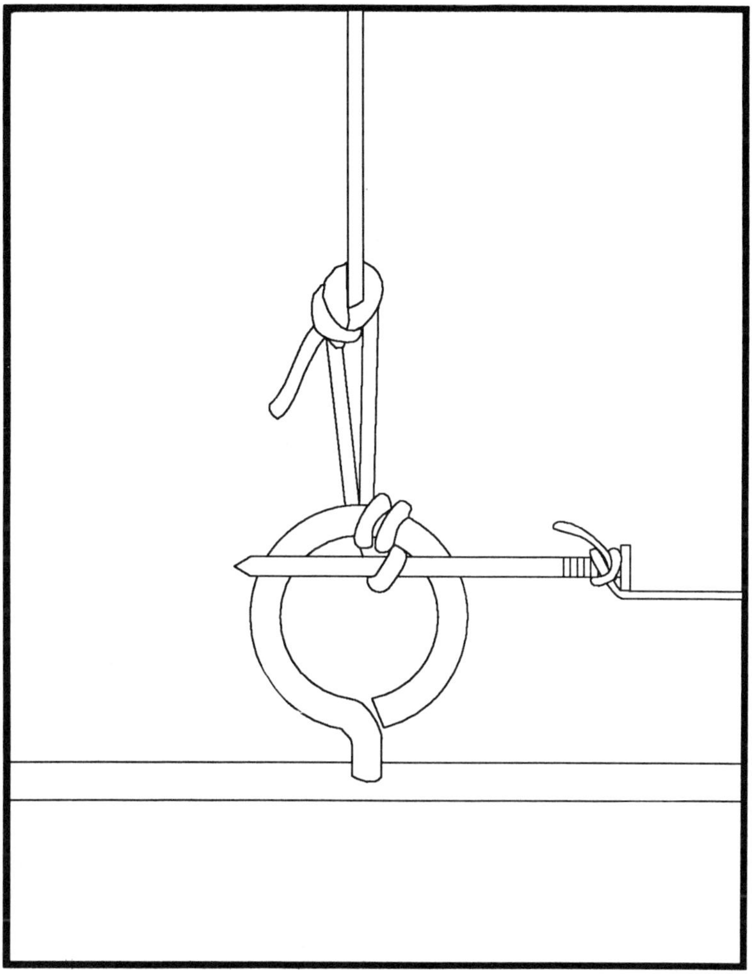

Fig. 5. Wrapping the line around the eye before passing the pin through it takes some pressure off the pin.

ses when only a partial wrap is taken, the force might cause the line to unwrap by turning the eye. If more than one wrap is made, it is importnat that one wrap does not cross another. Otherwise (remembering Murphy's Law) the line might find a way to tie itself in a knot. If an open eye is used (two parallel nails), there is a chance that the loop will be caught as it unwraps. Either of these occurances can render a trap useless.

While a wrap takes pressure off the pin, it does not always make the pin easier to pull out. When holding large amounts of weight a wrap can make a pin more difficult to pull out by locking it in place. The number of wraps required to make it easy to pull out will cause drag on the load line. If a hair trigger is required with a heavy payload, a pin and eye with wraps is not the best choice. Any bend in the load line can be considered a percent of a wrap. Two or more bends might equal or exceed one wrap.

Pin and Eye with Treadle (Pressure Plate)

Treadles have been used since ancient times to magnify the power of foot pressure. In this case they magnify the effective weight of the target foot, and so increase the sensitivity of the trigger. In addition to the power of leverage, the treadle adds its own weight to that of the target. The longer it is and the more distance there is between the target foot and the pivot (dead) end, the more power is magnified.

Rather than being kicked, as a tripline is, a treadle is stepped on. This makes it ideally suited for areas where the target will be descending onto a surface (basement window). It should be wide enough to form a platform that will not be missed.

The pin and eye are positioned over one end, with the pin pointing straight up and down. A line runs down to the treadle, and holds it suspended a few inches or so (depending on how much give the line has) above the floor. The other end rests on the floor or ground, or sits on a block. Pressure on the platform pulls the pin. (Fig. 6)

A rug can be laid over it for concealment; or it might be placed higher, and made to look like the top of a book case or something similar.

It is possible that the weight of the treadle will be enough to pull the pin by itself. This can be remedied by stuffing a few rags or other spongy material under it to take some of the weight. Or the line to the pin can be wrapped around a nail.

A pulley positioned over the floating end can turn the line so that release can be affected at the top anchor. This will allow the payload to fall with no rigging dragging behind. It will take a great deal more pressure to pull a pin at this point, so it might be necessary to lay weights on the floating end of the treadle.

A treadle, or pressure plate, can be hidden easily by being set in a pit. If it sits several inches or so below the surface, the target's weight will be magnified by falling, making triggering certain even without a hair trigger.

The load line can come all the way to the pressure plate, or a line can run from the pressure plate to another point. Either way, one line will usually need to run along the surface of the ground for some distance. It can sit in a shallow ditch, covered with leaves, or hidden in some other way. It should not be covered with dirt or anything else that will inhibit its movement.

A loose floor board makes a very good pressure plate. The pin and eye is set where the floor and wall come together, and the load line runs along the wall.

Treadles can be used as well with all of the trigger mechanisms discussed in this chapter, but their best use is with the pin and eye.

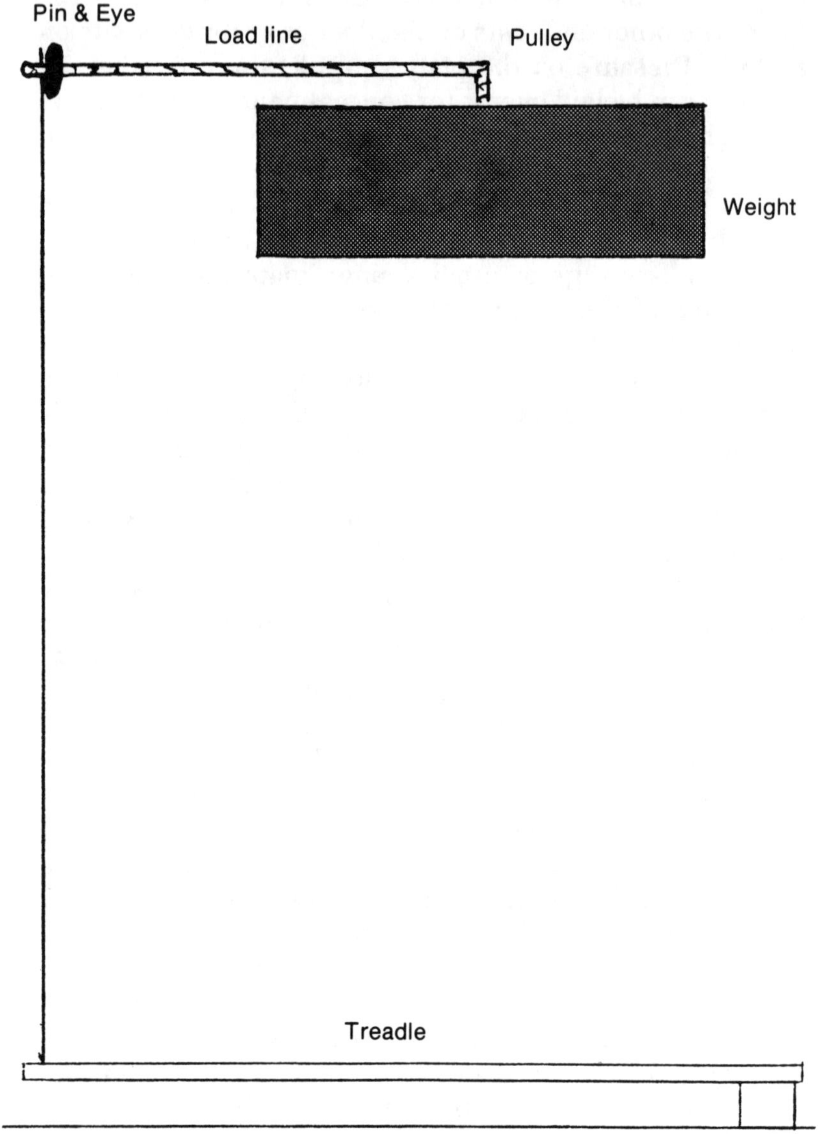

Fig. 6. Treadle activated P&E. Set in this way, the treadle does not magnify the force of the target's weight.

Ladder Trigger

A ladder rung can act as a treadle if one end is free and the other hinged. A line running up the inside or outside of a ladder leg should be more or less out of sight if it is in contact with the leg. If necessary it can sit in a shallow groove. Since the target will reach up to pull on the rung, the line will be above eye level. Triggering is certain even without a hair trigger, because the target will put weight on the lever.

This is an especially good trigger, because the target will be on a ladder when it is activated. Being up on a ladder should increase the amount of damage done, especially if fairly high up, or if more than one person is on the ladder. The loose rung itself might cause the target to slip and maybe fall, after which he will be in a good position to be hit with a payload.

Like all treadle-type triggers, a ladder rung can be used to activate any trigger mechanism, but works best with the pin and eye.

Stud and Eye

This is a more simple version of a pin and eye, which can work very well in some situations. Rather than passing through an eye, the loop in the end of the load line is simply slipped over a stud. A tripline, treadle, or other lever pushes or pulls it off.

Peg and Lever

This trigger is strong, dependable, and adaptable. It can be activated by a pull or push in a wide range of directions, or by being disturbed. It has an even wider range of forms than the pin and eye, and so is even more adaptable to existing structures and materials. A hair trigger is easy to set, no matter how much weight it holds. With this trigger, a thousand pound hanging weight could theoretically be released by a mouse sneezing on the trip-

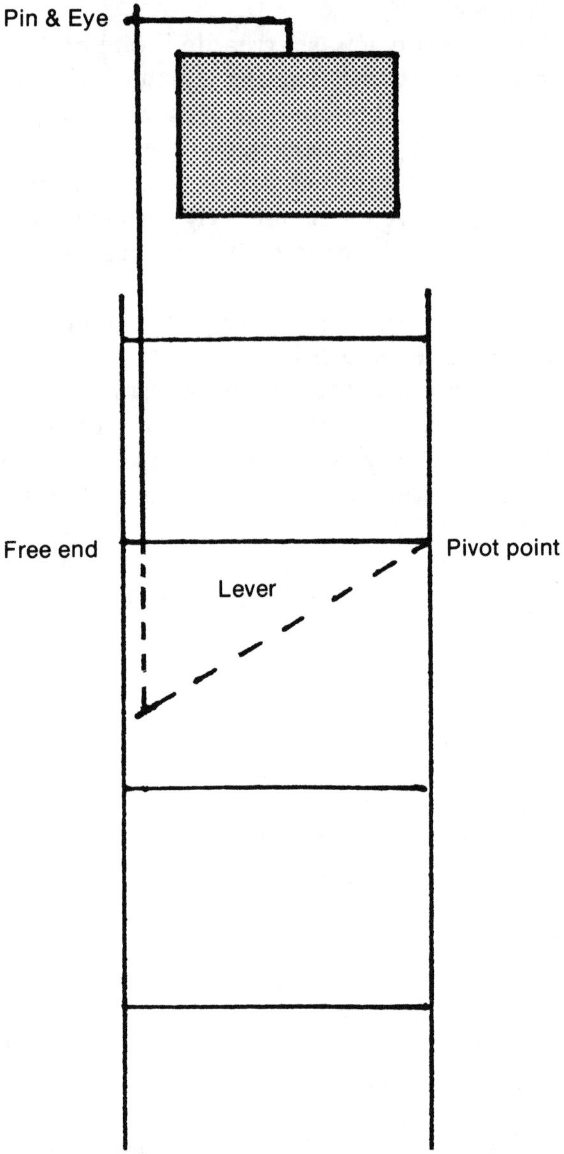

Fig. 7. Ladder trigger. A ladder rung with one end free can act as a treadle to pull a pin.

line. When the tripline is sewing thread and the payload is heavy, this should be the trigger mechanism of choice.

The peg and lever can be made from a variety of materials, with wood usually being the easiest to work with. Wood is almost always easy to get, in the form of trees, houses, furniture, etc.

The mechanism requires a beam that is out away from and more or less parallel to a surface, and two pegs. The lever peg needs to be cut to an approximate size, depending on the trigger's form. The holding peg needs to be a more specific size, which is slightly less than the distance between the beam and the surface.

To set the trigger, a large non-running loop is tied in the end of the load line. The lever is inserted part way through the loop, and passed between the beam and the surface, so that it is roughly parallel to the surface. (Some manuals show the line tied tight to the lever, but this is altogether unnecessary, and a peg attached to the end of a running load line can cause problems, for reasons discussed elsewhere) The holding peg is then fitted between the free end of the lever and the surface.

It can also be done with the load line rather than the lever passing between the beam and the surface. In either case, the lever holds the line, and the peg holds the lever. Removing the peg effects release. (Figs. 8 & 9)

If the distance between the beam and the peg end of the lever is greater than the distance between the beam and the surface, the lever will not pass through. This may or may not keep it from releasing, depending on the angle the load line is coming from. A little enough angle will allow the loop to slip off the end, providing nothing gets in the way. The amount of lever passing through the loop should be as little as possible, and free of obstructions, so that the loop can slip off easily. If the load line passes under the beam, the lever must pass through, no matter what the angle of pull is. (Fig. 10)

Fig. 8. Peg and lever assembly, with the lever passing between the beam and surface.

Fig. 9. P&L, with the load line passing between the beam and surface.

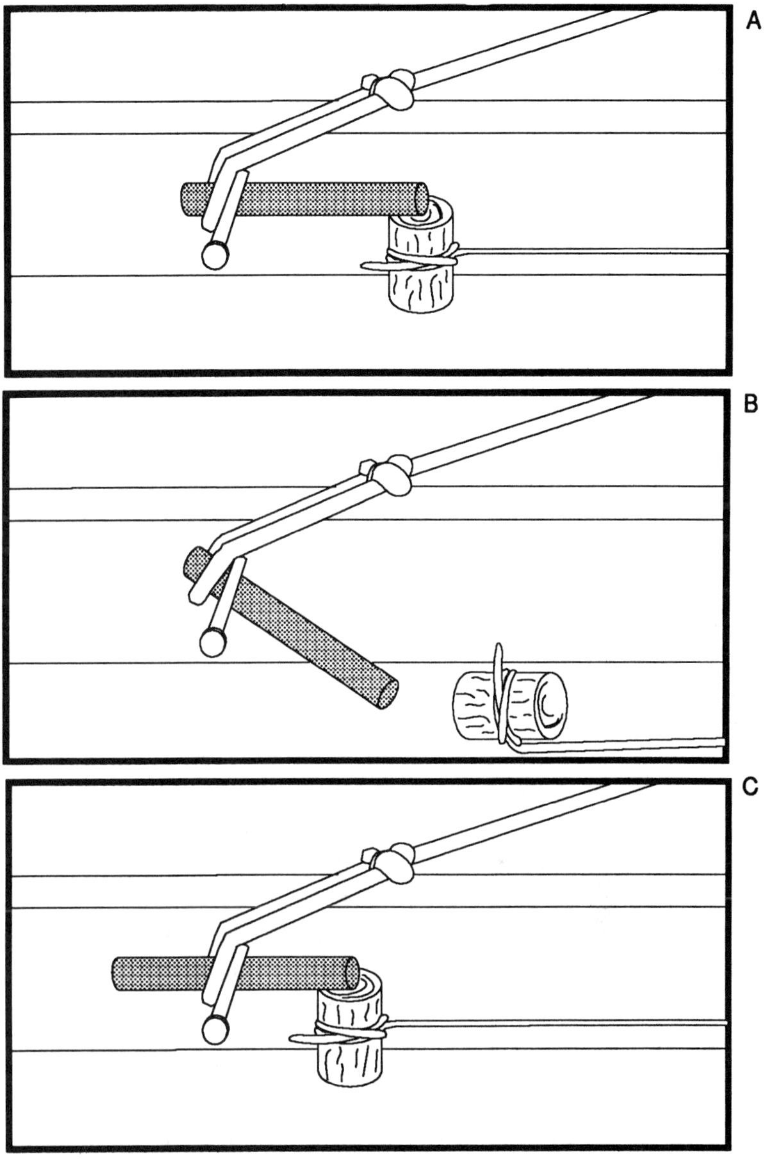

Fig. 10. The lever in (A) will not pass through, causing the mechanism to jam (B). (C) shows a correct setting that will not jam.

The load line and lever must be straight with each other, and both must be approximately perpendicular to both the beam and the surface. The lever must push straight into the peg, and the peg must point straight to the surface. If the angles are not correct, the structure will fold, and release. Sometimes another surface (such as a fence) can be used to keep it from folding. (Fig. 11) It can be shored-up in other ways as well, but doing so complicates the mechanism, and increases the possibility of malfunction. In some cases the load line will need to be turned in order to come in straight.

Because a lever is being used, the amount of pressure pushing against the peg can be reduced dramatically. The beam acts as a fulcrum, with the weight and the peg the opposing forces. The closer the force is to the beam, the less effect it will have at the other end of the lever. The further the peg is from the beam, the more effective strength it has. So, in order to get the maximum power from the lever, the load line should be as close as possible to the beam, and the peg all the way to the opposite end of the lever. (Fig. 12)

A hair trigger is set by touch. Just as it is possible to set a weight on a ledge so that it is easy to make it fall, the peg can be set so that the structure is almost ready to fold. If it is made too sensitive, it might trip prematurely.

The tripline can be tied to the peg anywhere, with the point where the peg meets the lever being generally best, as the weakest part of the structure. A pull in almost any direction will cause release, either by removing the peg, or by collapsing the structure.

A hard trigger attached securely to the peg can cause release if moved in any direction. But the mechanic must make certain that the trigger does not get in the way of the lever as it unwinds from the beam.

In some cases no tripline will be necessary.

If the peg and lever are not sufficiently strong to

Fig. 11. Since the load line does not come in straight, the peg and lever would fold if the fence did not help support it.

A

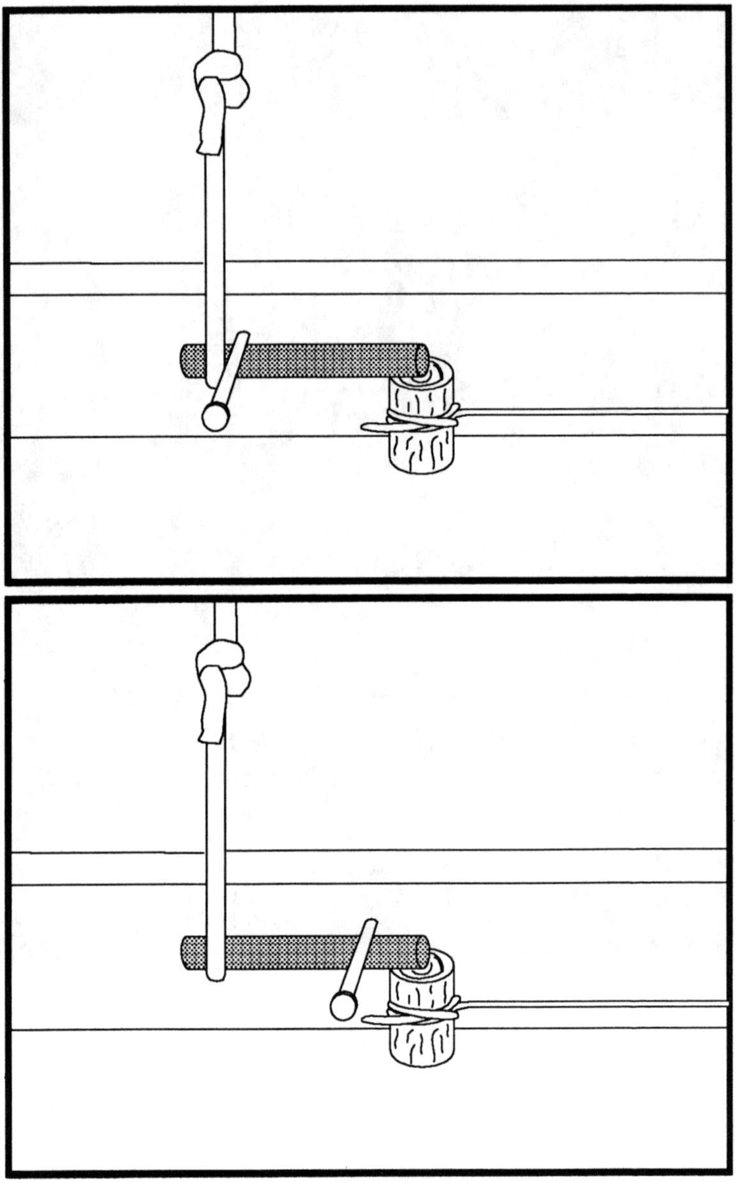

Fig. 12. Setting (a) puts the least pressure on the peg; (b) puts the most pressure on it.

support the weight, the line can be wrapped partway or entirely around the beam before the lever is inserted. If a full wrap is used, it must be done very carefully, because it can easily cause a malfunction. The lever will travel around the beam as the line unwraps. It must be short enough to do that. The wrap should be spaced away from the lever, so that the lever does not catch the load line. Because of the danger of malfunction, a wrap should not be used unless absolutely necessary. A hair trigger can be set easily enough without one. (Fig. 13)

A mechanic working or experimenting with a peg and lever trigger should be aware that the lever will usually be thrown with a good amount of velocity when the load line pulls free of it. Though small, this is a potential danger.

Curtain Trigger

A peg and lever can be set with a curtain between the peg and the surface that it pushes against. If other points on the curtain are anchored, the peg will be pulled out by the curtain being moved. Disturbing the curtain activates the trigger. The same curtain can hide all or part of the trap.

Pin Release Knot

This knot is held tied by a pin, it will hold a loadline securely to anything that the line can be wrapped around (i.e. nail). When the pin is pulled out, the knot releases. This leaves the line free of any knot, which is a big asset if the line will need to run through small pulleys, or if there is anything around that might catch a loop.

The end of the line is doubled back on itself. Then the loop end is passed around a stud and folded back again, so that there are four strands running parallel to each other. The pin is inserted between the loop strands and the other two, with the loop strands to the outside. The pin will be parallel to the stud. (Fig. 14)

A

B

C

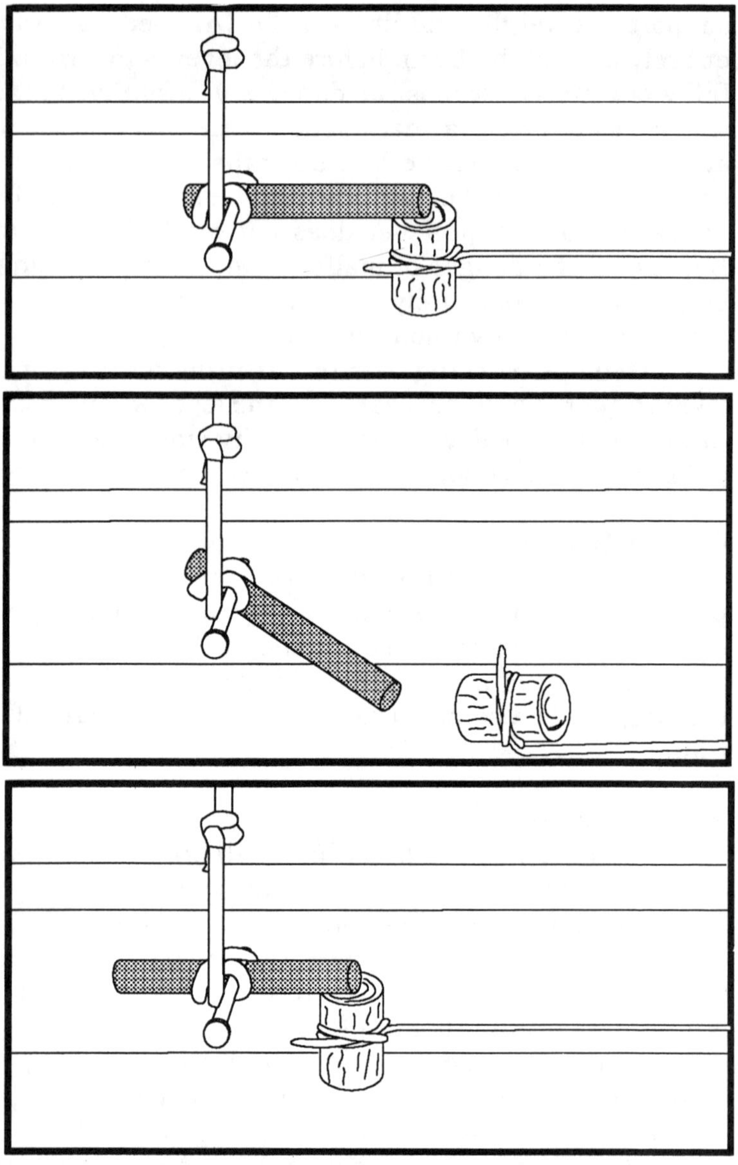

Fig. 13. A wrap takes tension off both the peg and the lever, but greatly increases the possibility of a malfunction, because the lever must travel all the way around the beam. Setting (a) will jam (b). (c) will not jam.

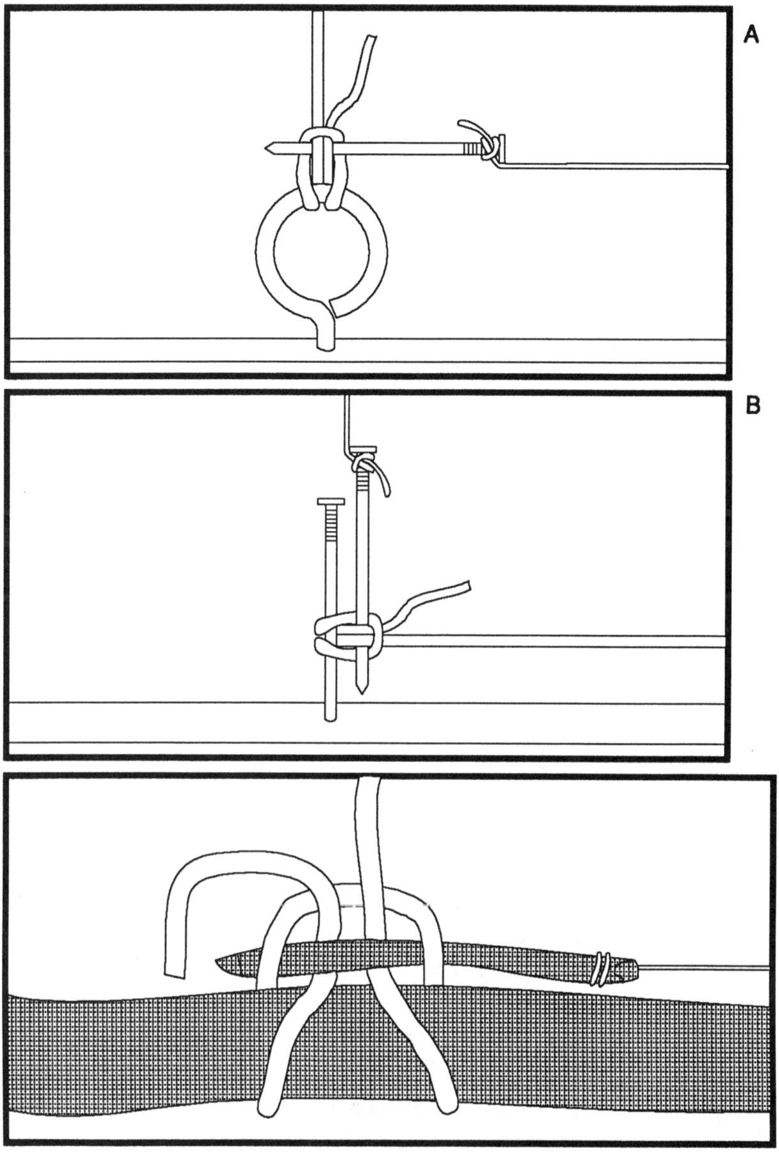

Fig. 14. When the pin is pulled out, the knot unties instantly, leaving the load line free of any knot. In (a) the load line must be approximately perpendicular to the stud; in (b) it can come in from any angle. In both cases the pin must be pulled straight out.

The load line must be approximately perpendicular to the stud. If there is too little angle, the knot will slide, and might slip off the end of the stud. If something like a screw eye is used, the line can come in from any angle.

The pin must be pulled straight out in order to release. It might be necessary to turn the tripline. This trigger is used in about the same way as the pin and eye.

Vice Grips

As mentioned in the Introduction, this is one of the most useful of all boobytrapping tools. In addition to taking the place of several less versatile tools, it can be used as a trigger. It is not as good as other triggers, and using it will usually mean sacrificing it. But if nothing else is available, it will work.

The load line is clamped to anything strong enough to hold the weight. The jaws can be adjusted to fit a good range of objects.

A hair trigger is set by adjusting the jaws so that they are almost ready to spring open. This will involve some trial and error. If a lot of weight is used, a hair trigger will be difficult or impossible to set.

It is important that the release handle be pulled straight out. Otherwise the jaws will never open. It might be necessary to turn the tripline. (Fig. 15)

A long stick attached to the release handle makes a good trigger. By acting as a lever, it reduces the amount of pressure needed to open the jaws, making the trigger much more sensitive. The stick does not need to follow the line of the handle exactly, and so can be angled so that it is in the right place in relation to the payload. If there is insufficient time or material to attach the stick to the handle, it can be placed between the handles, with the end against something (the stud that the jaws are clamped on, for example). When the lever is pushed, it will pry outward on the handle.

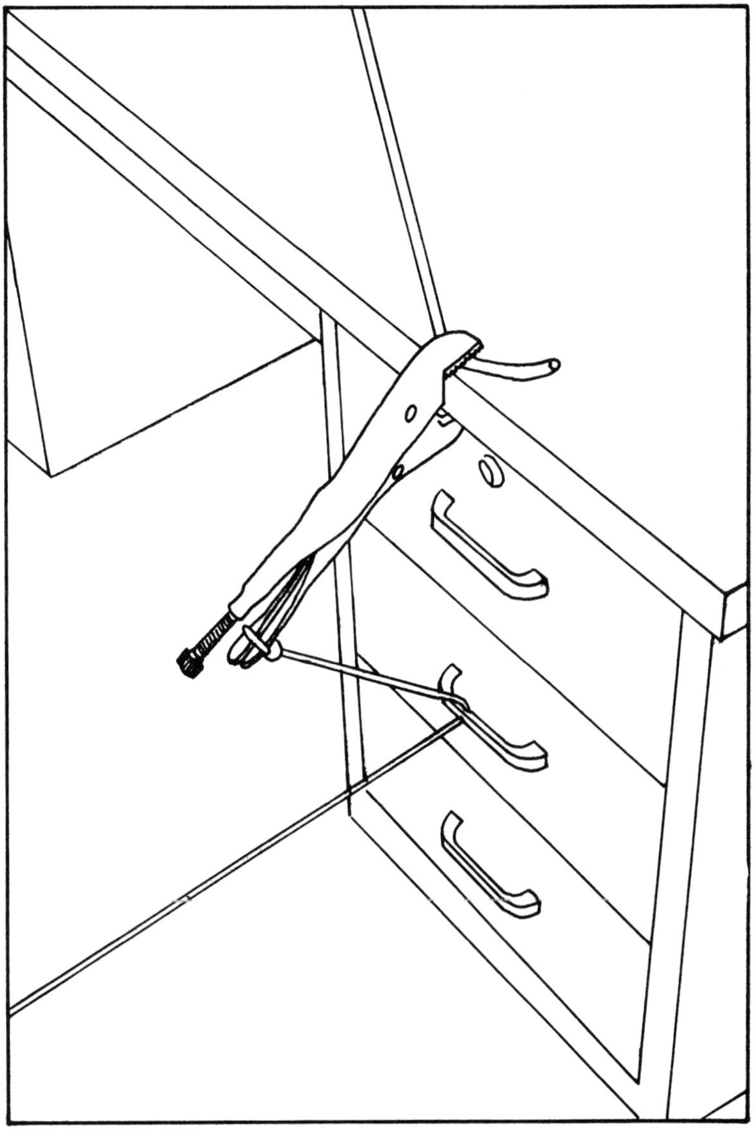

Fig. 15. A pulley is used to turn the tripline, so that the handle can be pulled straight out. If the drawer opens too easily, the trigger will fail. Note that this trigger has the same feature that the PRK has: the line is left free of any knot.

Balance

This is the simplest and often fastest way for this type of trap to be triggered. All that is required is a ledge for the weight to sit on. It must be able to support the weight, and it must be in the right place. It can be a ceiling beam, a shelf, a staircase, furniture, roof, or a tree limb.

The weight is moved back and forth on the ledge until it is almost ready to fall off. It can be tripped either by being disturbed, or by the surface it sits on being moved.

The classic example of the latter method is a weight placed between the top of a barely open door and the trim around the door frame. Since the ledge will be higher than the top of the door, it might be necessary to place a spacer between the top of the door and the weight (Fig. 16). In either case, the weight must push straight down on the door, or it will push it open and fall. If there is no ledge around the frame, one can be made by hammering in something like a nail or a knife.

If the weight is pulled down by a tripline, the line should be attached by a large open loop, so that the weight will slip free of it as it falls. In some cases this will not be necessary.

Miscellaneous Triggers

Obviously, there are many ways to trigger deadfalls. The above list covers the methods that are generally best, but in some cases a different method may be preferable. It should be remembered that the principles covered at the beginning of this chapter should apply to any trigger.

As discussed above, wrapping the load line around something takes tension off of it. Each wrap will take more weight, and if enough wraps are taken, the load line will be secured just as well as if a knot were tied to it. So it is possible to take just enough wraps that the line is almost secured, but not quite. With so little tension on it, it can be held by methods that would not otherwise work.

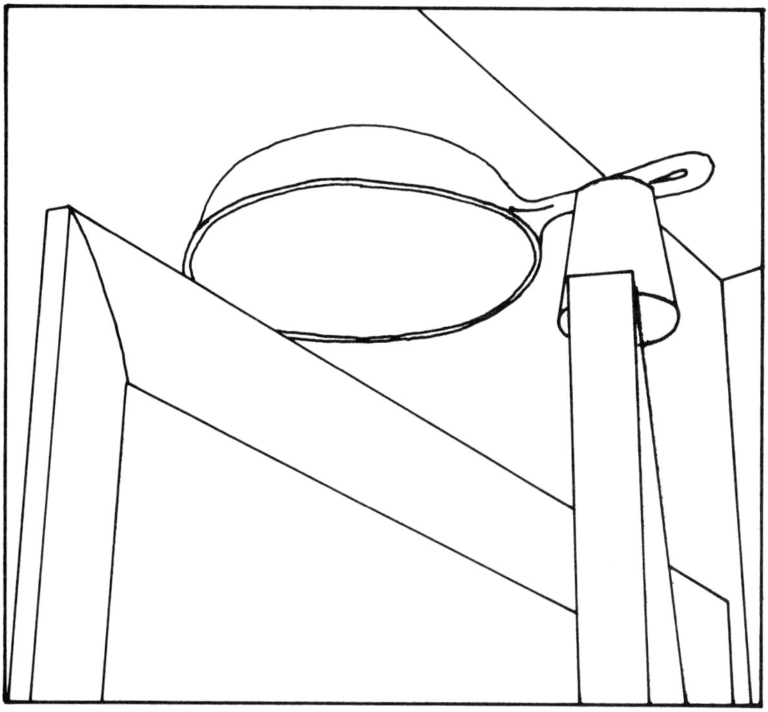

Fig. 16. This is the fastest and easiest way to make a straight deadfall, yet it can be very effective. Note cup used as spacer.

The amount of line on the free side of the wrapping should be no longer than is necessary. All of it will have to follow the course of each wrap, and so will put drag on the payload until it unwraps completely. This will delay the fall slightly, and will rob power.

A weight can be set on the line so that it holds it until it is moved (i.e. a vehicle with one wheel sitting on the line). This may be accompanied by wrapping. If the weight is not quite sufficient to hold the line, a knot can be tied in the end of the line to keep it from sliding under the weight. If the weight is not something that the target will intentionally move, and it is too great to be tripped easily, a lever can be used to magnify whatever pressure the target puts on it. The lever may or may not be the trip line. It might be used as a treadle.

Another adaptation of the above principle is to close the line in a door, window, drawer, or anything else that will be opened by the target. Wrapping and knots might be used.

Counterweight

A payload can be hung with an equal or greater weight at the opposite end of the load line acting as a counterweight. When the counterweight is released, the trap is activated. A good example of this is a bound or dead comrade of the target hanging by a rope from a tree. The rope disappears into foliage. There it runs over a branch (pulley) and is attached to a payload. When the bait is cut loose, the payload falls.

Climbing Rope

A rope that appears to be used for climbing can be attached directly to a payload. When a target puts weight on the rope, he pulls the payload down on top of himself. People usually test climbing ropes by pulling on them before putting their weight on them, and usually look up and are ready to move in case the rope comes loose and falls. But not always. If a person is in a hurry he is less likely to test a rope, and if he has just seen another person climb the same rope he is very unlikely to test it. So a good way for this trigger to be used is for a person being chased to climb the rope, then immediately untie it from its stable anchor and tie it to a payload. The trap can be set beforehand.

Fire Release Trigger

A payload can be hung with burnable rigging so that fire will cause release. Several fuel containers with triggers of this type can be used together with one that has a target-initiated trigger. In this way a trap can contain a large amount of fuel, and can be spilled at different points (i.e. at each exit), but only one complex trigger needs to be made.

Tension Release Trigger

A dropping weight can be used to make a trap capable of being activated either by tension or tension release. In that way if a tripline is discovered and cut, the trap will still be activated.

The weight sits balanced on a ledge that will allow it to fall either way (i.e. a nail sticking out of a wall), leaning slightly away from the tripline. The tripline holds it from falling, and it holds the tripline tight. Pressure on the tripline will make it fall one way; release of the tripline will make it fall the other way.

A line leads from the weight to a secondary trigger assembly. This can be a pin and eye, peg and lever, or any other mechanism. (Fig. 17)

Multiple Triggers

One important thing to remember in regard to triggering is that action can always produce more action. One

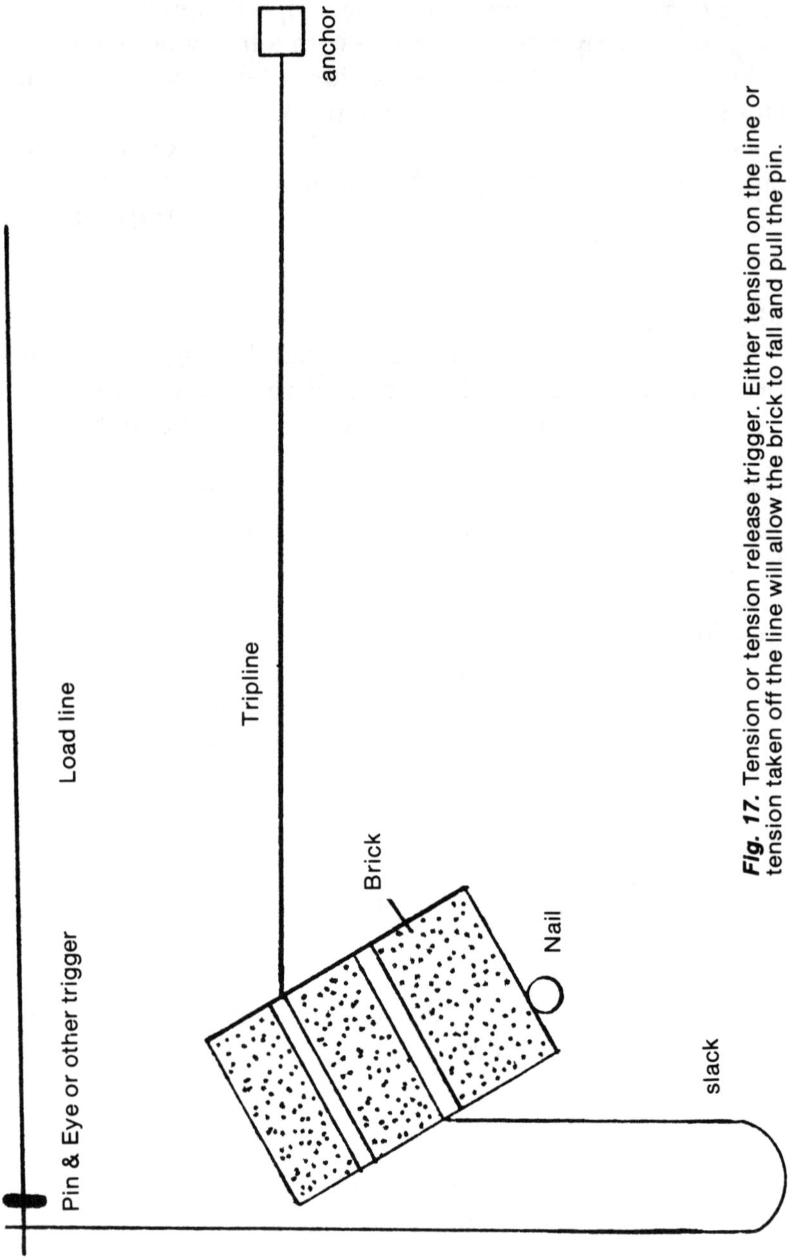

Fig. 17. Tension or tension release trigger. Either tension on the line or tension taken off the line will allow the brick to fall and pull the pin.

trigger can activate another. Just as a small electric current is channelled through various mediums and ultimately becomes a massive explosion, a very small trigger can drop any weight. Using fire as a medium, the smallest igniter discussed in this book can ignite any amount of fuel. A dropping weight can pull any number of triggers. This may be in addition to its function as a payload.

CHAPTER 4

Rigging

Rigging is the trap's energy system. It consists of a load line and one or more pulleys.

The load line runs from the payload through the overhead pulley to the trigger. It is attached there to the trigger assembly, and holds the payload poised over the target area until released. It must be strong enough to support the weight of the payload, but otherwise should be as small and light as possible, so as to offer a minimum resistance to the payload's freefall (drag). Braided nylon cord is the best load line material for most purposes. Since smallness of the load line is an asset, lines should be used to near the limits of their strength.

Action between the load line and pulley should be as smooth as possible, so as to keep drag to a minimum. The pulley should be substantially larger than the line, and have a smooth surface. The line should be free of obstructions (i.e. knots) that will not pass easily through pulleys (Fig. 18).

Usually the line will need to be turned in order to be kept out of the target's field of vision (Fig. 19). This will require the use of one or more additional pulleys. The

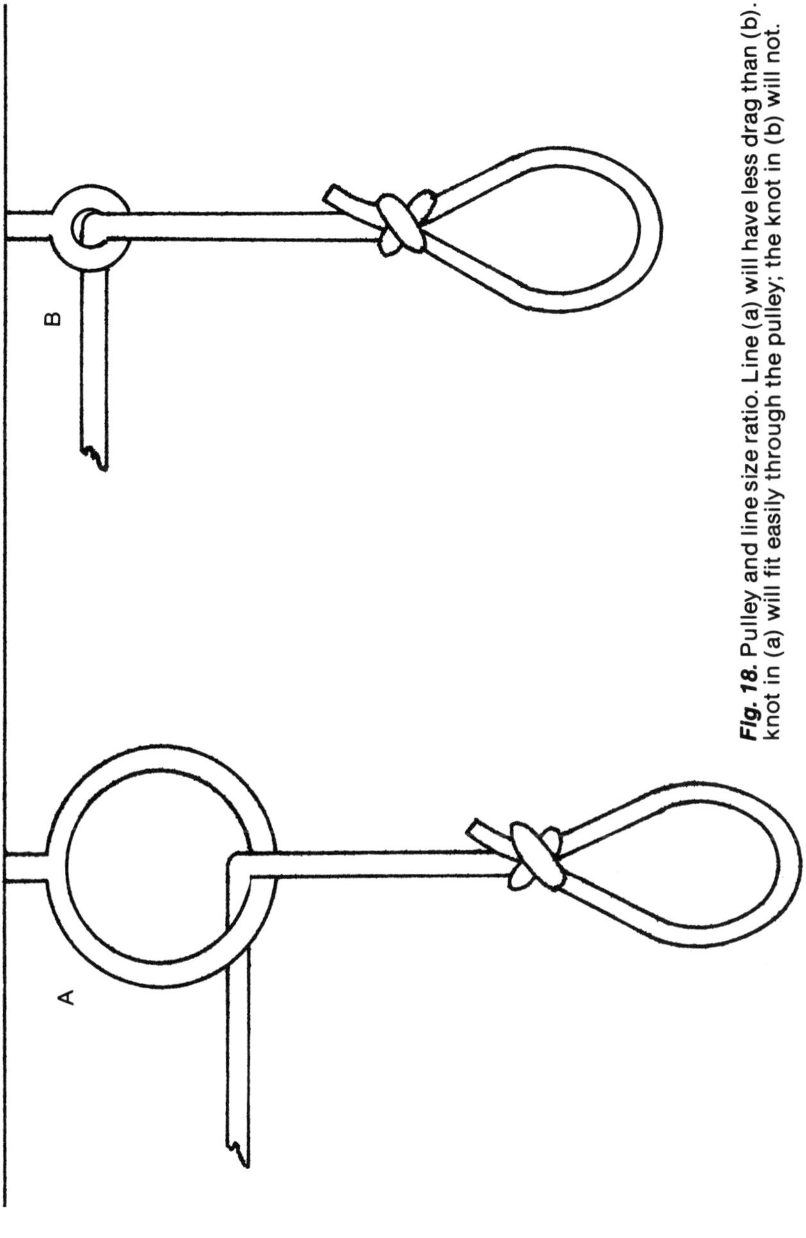

Fig. 18. Pulley and line size ratio. Line (a) will have less drag than (b). The knot in (a) will fit easily through the pulley; the knot in (b) will not.

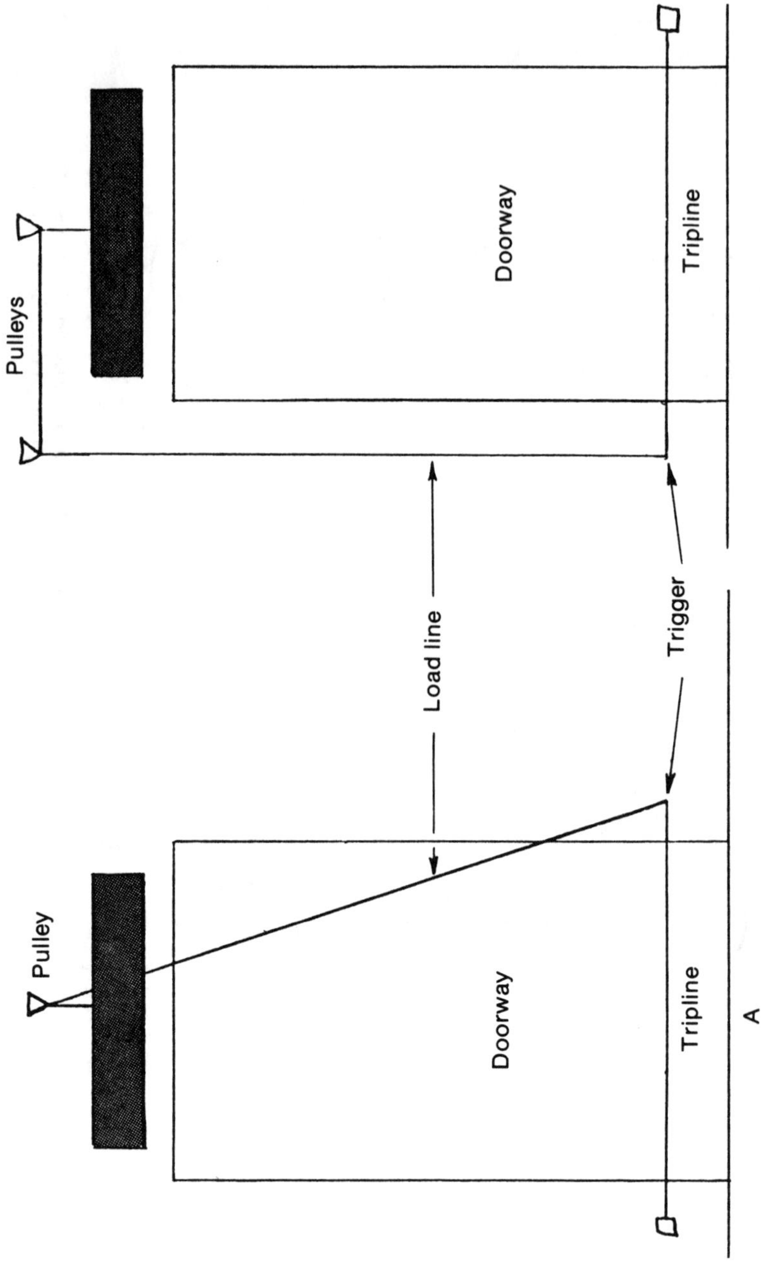

Fig. 19. The load line in (b) is turned to keep it out of the target's field of vision. The line in (a) is visible and conspicuous.

number of pulleys should be kept to a minimum, and the line should be bent no more than necessary. The straighter the line between the payload and trigger is, the less it will drag.

Unless the line is overly large, or rigging is done very poorly, drag will not be a significant factor, and can be ignored when making time and energy calculations.

A pulley is anything that will hold the line in a certain position, and at the same time let it run. This can be factory-made pulleys as sold in hardware stores, a screw eye, fencing staple, nail (which may or may not be bent), and so on. Holes in boards such as wall studs, ceilings, etc., crotches where tree limbs come together, and similar items make good pulleys. A pulley that does not form a closed circle (i.e. nail) might catch a running line that has a loop in the end, causing the mechanism to fail.

The load line and trigger can come together at any pulley, including the overhead pulley. Release is effected at this point. Release at the overhead pulley will reduce drag to zero, but will usually complicate triggering.

Aim

Position of the overhead pulley determines where the payload hangs, and so where it falls. It must be aligned directly over the intended target area. If falling distance is short, correct alignment is easy to arrive at using sight alone. Greater distance requires more precise setting. This can be done in two ways. One is to let the payload hang a few inches off the ground and note its position before it is hoisted. The other way is to use a plumb bob (see Chapter 1).

A payload will usually be set to cover the same area covered by the trigger, but not always. For example, a trap might be set to pour burning fuel out at the only entrance to a room, but be triggered from deep inside the room. In this way multiple targets can be hit by a single trap.

Attaching Lines to Payloads

The load line must be attached to the payload so as to allow the payload to hang correctly. The payload's center of gravity (CG) will always hang directly under the line, no matter how or where the line is attached. So to get a payload to hang correctly, the load line is attached at one or more points that will put it in line with the payload's center of gravity when in the correct position. For more information on center of gravity, see Chapter 11.

The line will be tied directly to the payload itself, or to one or more brackets on the payload.Brackets must be placed so as to allow the load line to be over the CG (Fig. 20). When more than one bracket is used, a line is tied between them, and the load line attaches to it with a non-running loop so that it can easily be adjusted to find center.

Slings

The most common way for things to be lifted on construction sites is with slings. These are either lines with eyes (non-running loops) in either end, or with the ends coming together so that they form a circle. These can easily be made by braiding or tying an eye in one or both ends of a section of cord, or by simply tying the ends together. They can be used in a variety of ways.

Slings are often better than knots for attaching lines to loads, as well as easier. As long as there is weight on the line the sling grips tight and holds the load securely; when the line does not have weight on it the sling can easily be loosened and moved. Often it is necessary to raise and set down a load several times to find exact center so that it will hang correctly. Each time the load line must be adjusted toward the low end. When it is attached with a sling, this is easy to do.

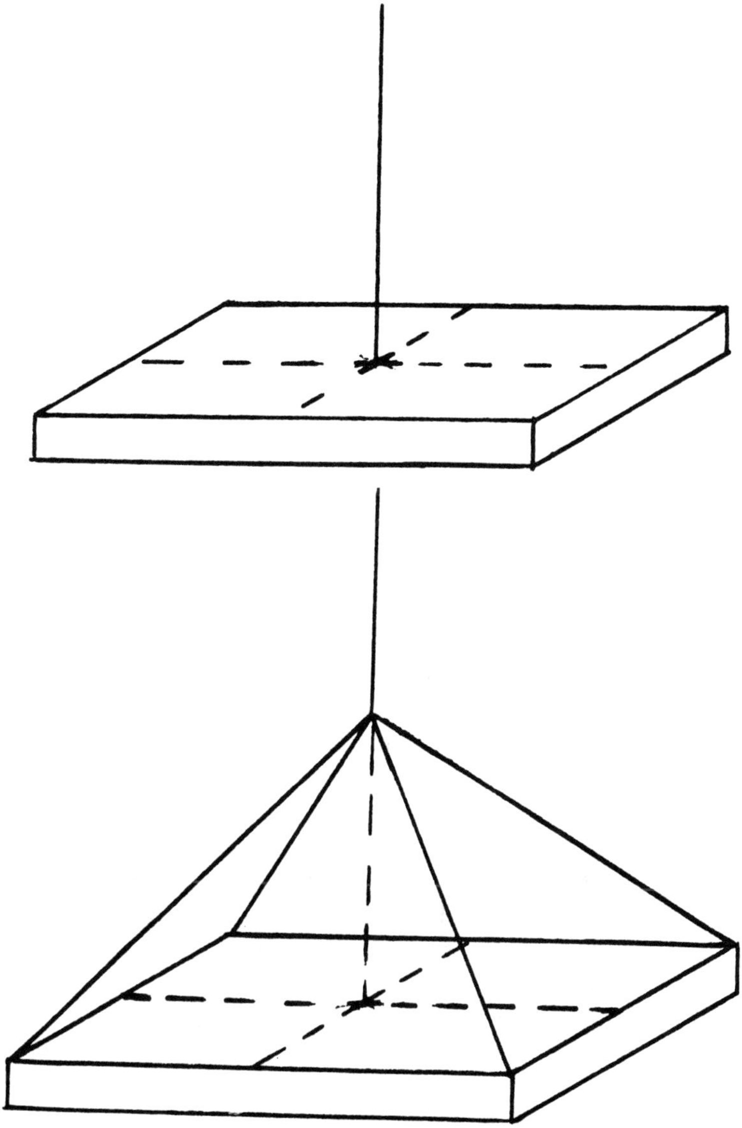

Fig. 20. The load line must be centered both ways, regardless of the number of points in attaches.

CHAPTER 5

Payload

The body of the trap is its payload. It is what contacts the target and delivers the trap's kinetic energy into it. It can have many forms, ranging from primitive to sophisticated, and simple to complex. Sophisticated payloads are potentially more powerful than more primitive forms, and are often faster and easier to build. But they tend to require a great deal more precision, and so are more difficult to build correctly. Primitive forms can be made from a greater variety of materials, and so are more commonly available.

Primitive Forms

Weight alone is the most primitive form of payload. Weight coupled with a device to concentrate it (i.e. spike) is one step more sophisticated. Weight is the origin of the payload's power. Velocity (falling) and concentration (spikes) magnify power, but it begins with weight. More weight means more power. Overkill in regard to weight is highly recommended, up to the point where rigging becomes unnecessarily difficult.

Energy as expressed in foot pounds can easily be

calculated: weight times falling distance equals impact force. (This should not be confused with momentum, which is calculated in a different way.) Pounds times feet equals foot-pounds. Power is increased by raising either quantity. A one pound weight falling one foot will hit with a force of one foot-pound; ten pounds falling ten feet will hit with one-hundred foot-pounds; a four pound hammer falling five-hundred feet will hit with a force of one ton.

The effect any given energy has on a target is extremely variable. It depends on the size of the area that contacts the target (i.e. how sharp the spike is), the softness (elasticity) of the target, vulnerability of the target, hardness of the payload contact surface, and whether or not the target absorbs all of the force.

Weight, falling distance, hardness, and sharpness all go together to produce killing power. Increasing any factor will increase power.

Concentrated weight is the best and easiest to use. This includes most forms of steel, such as beams, train rails, barbell weights, cooking pots, vehicle parts, manhole covers, nuts, bolts, nails, and so on. Silver, gold, and copper all weigh more than steel, so things like jewelry and coin currency contained in a vessel will weigh more than an equal amount of nuts and bolts. Lead, in the form of bullets, shot, and so on is a great deal heavier than steel. Bricks, rocks, and sacks or boxes filled with sand or dirt are good.

Clay is good because it can be molded, and it is almost always easy to get. In many areas it can be dug out of the ground. It can usually be found along river banks, and areas where large bodies of water have dried. Ground that forms cracks when it dries is likely to be all or part clay.

Clay should be mixed thoroughly and kneaded until it is a uniform, putty-like consistency. After being molded it is allowed to dry slowly. It can then be hardened further by being fired. One way of doing this is to build a fire on top of it. For more information on working with clay, consult books on clay modeling, primitive brick making, and pottery making.

Anything that can be done with clay can usually be done better with cement.

Spikes magnify power by concentrating force onto a small surface. This allows the weight of the payload to penetrate easily into the target body. Their primary requirements are sharpness, hardness, strength, the ability to be attached to a payload, and suitable length. Nails, and tools such as knives, ice picks, and screw drivers are usually the easiest to get and use. Large screws and bolts can be filed or ground sharp, and are easy to attach to things. Wood, in the form of sharpened sticks, pencils, etc., can be effective. Any wood can be both hardened and strengthened by being saturated with something like super glue. Singeing the point on a stick in fire will harden it slightly.

If nothing better is available, broken glass can be used. This is not as sharp as is often believed, but can be effective if driven by sufficient force. Thick glass driven straight into a target can have a good amount of strength. If it breaks after entering the target body, so much the better. The shorter it is, and the straighter it drives, the less chance it will have of breaking instead of penetrating.

The more spikes there are, the more points on the target will be attacked. But since their purpose is to reduce the size of the area that contacts the target, too many are counter-productive. If enough can be known about the shape and position of the target, one is enough. As more weight is added, more spikes can be used, and so a larger area covered.

Spikes should be set so as to drive as straight as possible, or in other words, be set parallel to payload trajectory. This will increase their ability to penetrate a great deal. The shorter they are, the less important perfect straightness is. Sufficient overkill in regard to weight will make straightness unnecessary, because they will be driven in no matter how they hit.

Spikes can be made a good deal more deadly by being coated with a toxic substance. The classic example of such a substance is animal defecation (preferably human). This introduces disease-causing organisms into the blood. It is a very primitive form of biological warfare. Strong, concentrated poisons such as pure nicotine can insure immediate death, even if a vital target is not hit directly. When such poisons are used, they become the payload, and the spike a part of the delivery system.

Spikes coated with anything liquid are likely to rust, which will make them less sharp. Whether or not this is a problem depends on the nature of the payload. Since only the tip needs to be sharp, it can be covered with oil or grease, and the rest of the spike covered with the contaminant. Rust is a good carrier of micro-organisms.

PAYLOAD EXAMPLES

Vertical Weight

Weight that hangs and falls vertically has all of its substance behind one point, and so is used efficiently. Long slender objects used in this way will have a great deal more power than if used horizontally. They will also cover a great deal less area, and so are much less likely to hit a target. This type of payload should be used only when the exact position of the target can be known. It works well in combination with a pit.

Providing there is enough head room, this type can be easier to hide than other payloads, because it requires less space on the trip down. That allows it to slip easily and quietly through foliage and other things that might be between it and the target.

Objects that make good vertical payloads include logs,

steel and wooden fence posts, railroad ties and rails, angle iron, I-beams and other types of construction steel, pipes, rods, and so on. These can all have considerable power.

Since they require a lot of head room, vertical payloads are well suited to tall treetops, large buildings, construction sites, power plants and other industrial areas.

Since being vertical is what gives this payload its power, it should be exactly vertical when it falls. This means that it must hang exactly vertical. To hang vertically, the load line is attached as near to the top as possible. Ideally, the line should come from the center of the top, but this is not always possible.

A spike placed at the low end will of course concentrate power. It should be attached securely so as to penetrate into the target rather than slide up the payload. It can be wired, taped, glued, nailed, or attached in some other way. A horizontal ledge on the weight can be used to push against it. A screw driver can be set in the end of a pipe, and then the pipe hammered down to keep it from pushing in further. Or the weight and spike can be one and the same. A log can be sharpened and the end hardened in fire. Steel can be cut to a point with a torch, grinder, welder, or saw. Swords and spears are ready made weight and spike combinations.

Horizontal Weight

In most cases a payload should cover as much area as possible, which means hanging and falling horizontally. Logs, boards, and many forms of steel can be hung in this way. This type of payload works especially well with spikes; they concentrate the force that would otherwise be distributed by the horizontal position.

The load line can be attached either exactly in the center (in regard to weight), or at from two to four points equi-distant from center. Multiple point attachments tend to be more stable and easier to work with, because it is often difficult to find center from one point.

A load hanging from a single point (regarless of the number of points that originally lead off the load) will usually want to rotate somewhat, due to wind, etc. A payload that has to fall in a certain way, such as a log hanging over a trail or a road block, must be held in line with its intended target area. This requires some sort of guide, which will keep it in position, but not interfere with payload descent. The guide can be something like a wall or viaduct. To hold something like a log in line with a trail, two guides are placed, one on either side of the log center point. They should originate from overhead, so as not to interfere with the fall. They can be made from small branches lashed to one or more overhead limbs. In this way they can aid concealment as well as hold the payload in place.

Spiked Board

Almost any flat board can be used to hold spikes securely and support weight. It should be thick enough to hold the spikes straight and keep them from being pushed upward, and strong enough to hold the weight. Plywood is the best, because it will not split. Anything made out of flat lumber (walls, shelves, furniture, doors) can be torn apart and used as a platform. Small boards can be built into a platform.

Nails make excellent spikes, and are easy to get and use. Very long nails are available at most hardware and lumber stores. Knives, ice picks, etc., can be hammered through just like nails, but might cause some wood to split. Holes can be bored in a platform and a handle slipped up through, so that the platform pushes against the finger guard. Broken bottles can be glued on; the glass should be short and thick.

Spikes should form a tight enough pattern to insure that

the target (top of head, shoulder, etc.) can not fit between them. They should be placed as close as possible to the four sides, to maximize the area covered by the payload. If the board is the type that can be easily split, the pattern should not cause nails to be in line with the grain. Small splits can be repaired with glue, or by wire wrapped tight around the board across the grain.

Anything that has a substantial amount of weight can be set on the platform. Lashing the platform and weight together may or may not be necessary. Concentrated weight (steel, bricks, etc.) is best. A railroad tie or log can be nailed or lashed to a platform. A platform can be turned into a box by building side walls on it, and the box filled with weight such as steel tools, rocks, and so on; empty spaces between weights can be filled in with sand, dirt, etc. A heavy cardboard box can be set on top and similarly filled; its sides may need to be re-enforced with something like tape. Heavy bags (i.e. trash bags) can be used in the same way. Bags that are already filled, such as flour, fertilizer, etc., work well and save time. Figs 21 & 22

The load line can be attached at from one to four points. A sling can be slipped over the payload and used as a two point attachment. Two slings perpendicular to each other make a four point attachment. In either case the load line should be attached to the slings with an open loop so that it is easy to find center. Bent nails, screw eyes, bolts, and other brackets can be set either one in the center, at either end, or in each corner. Holes in the platform can be brackets. Fig. 23

A spiked platform can be made using only clay, a cardboard box, and nails or other spikes. The bottom inch or so of the box is filled with clay. Spikes are pushed through from the top. Then the rest of the box is filled and the clay dried.

This can also be done using cement instead of clay,

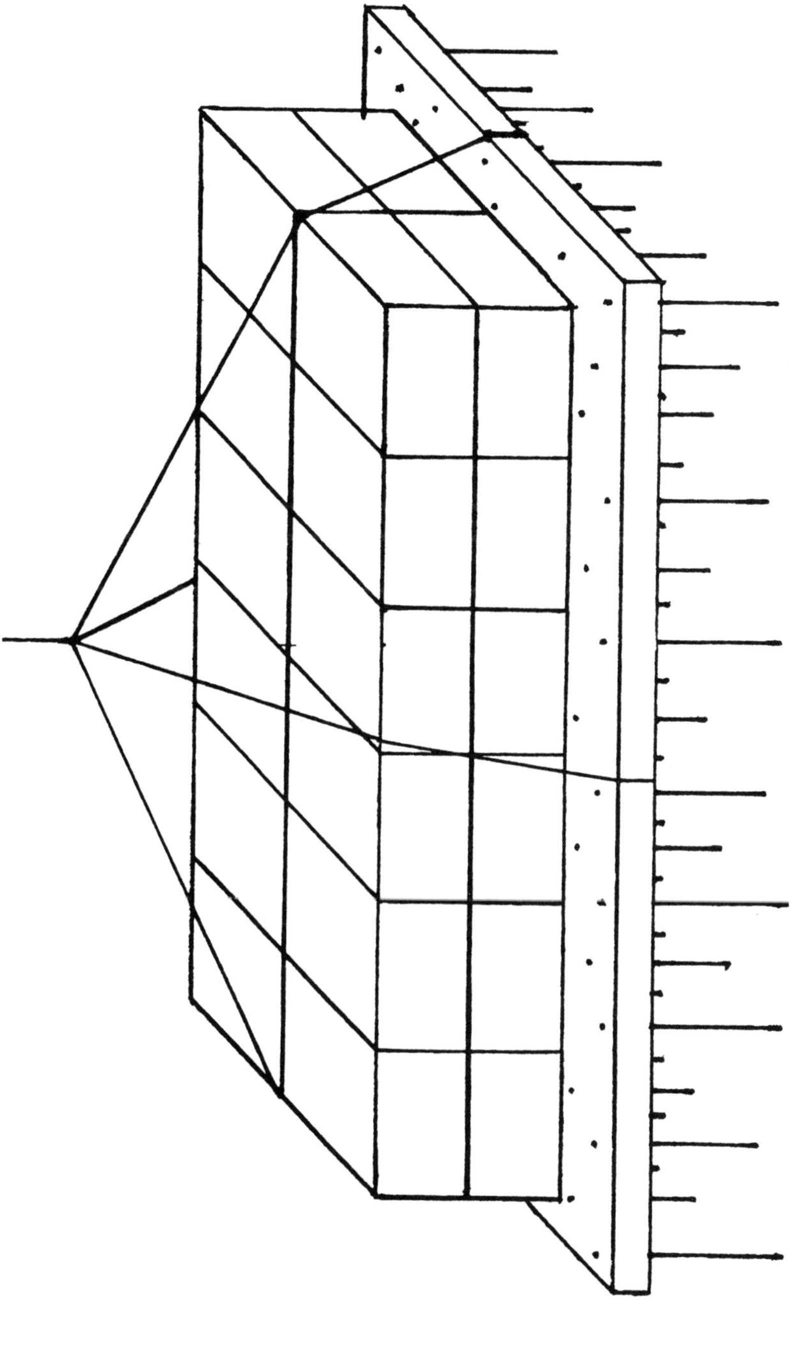

Fig. 21. Nail board with bricks stacked on top. The bricks do not need to be lashed down. It is important that the support lines do not fit between bricks.

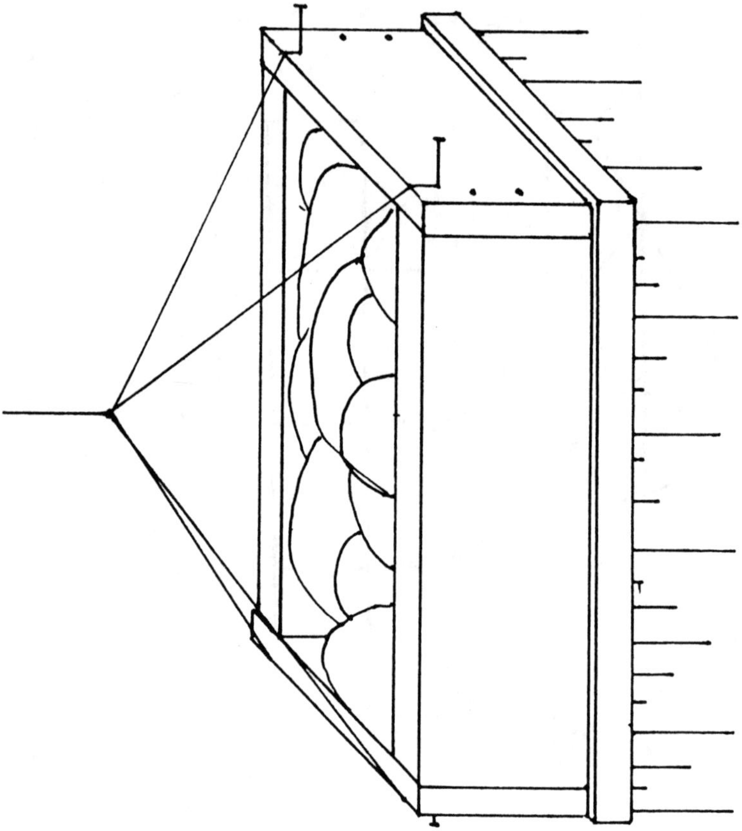

Fig. 22. Box nail board. The box is built on top of the nail board, and filled with the heaviest weights available. Empty spaces between weights are filled with dirt. Nail brackets simplify rigging.

Fig. 23. Brackets for attaching lines to boards.

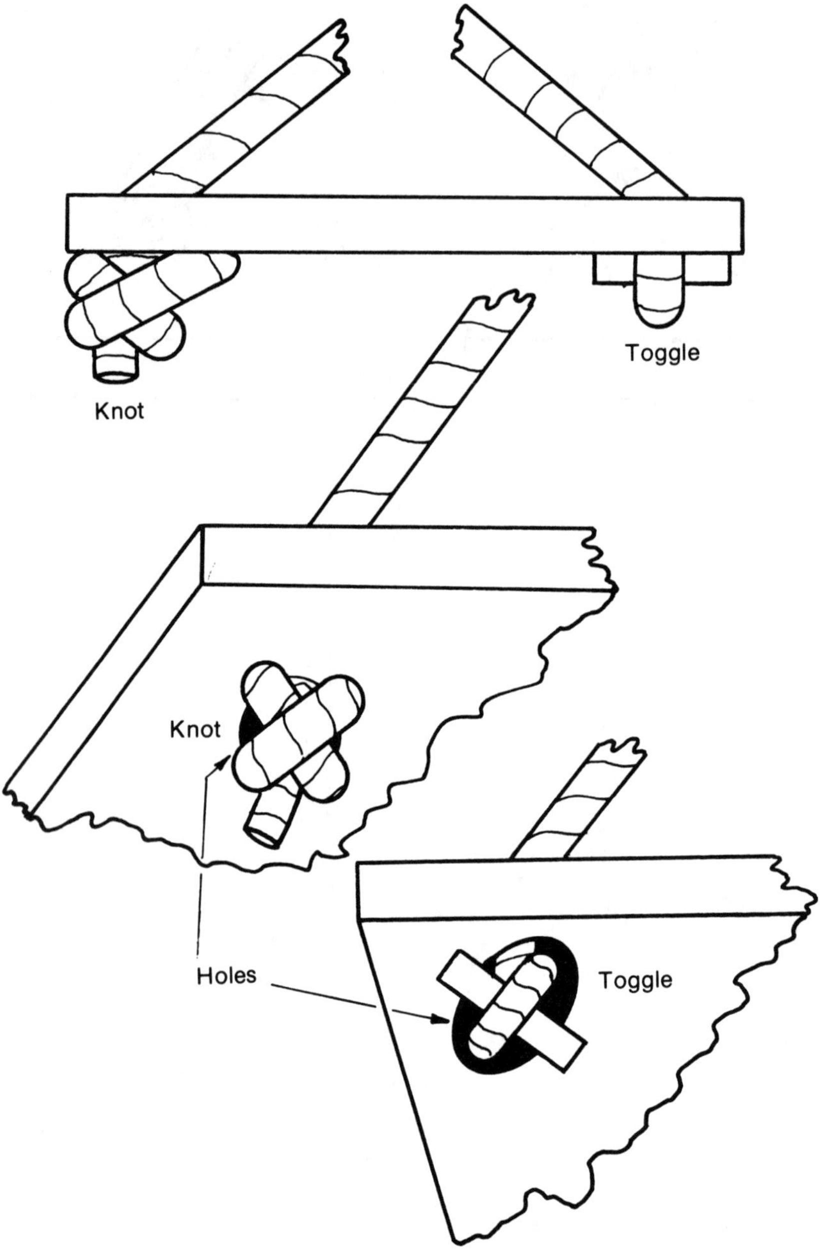
Toggle
Knot
Knot
Holes
Toggle

providing the box has sufficient strength to hold it until dry. Lining the box with plastic (i.e. trash bags) will keep it from soaking through and becoming weak. The box can be strengthened by being surrounded by a frame of some sort.

Logs and Rails

A log hanging parallel over a trail uses its size and shape efficiently to cover the greatest area possible. This simplifies timing, and can result in multiple hits.

The load line can attach at one or two points. Either way it should be centered so that the log hangs level (flat) over the trail. Falling flat makes maximum use of the log's power.

Some sort of guide should be used to hold the log parallel to the trail. The same guides can aid concealment.

A railroad rail can be used in the same way as a log, but will have more power.

Road Block

A large heavy object dropped onto a road will stop a slow moving vehicle or wreck a fast one. Ideally it should be long enough to cover the entire road, so that it can not be steered around. It should be at a point in the road where the vehicle must stay on the road, such as a bridge, tunnel, or overpass.

Many locations of these types offer good supports for overhead pulleys, and obstacles to hide trap parts. Fig. 24

Suitable payloads include logs, railroad ties and rails, or boulders. Anything large and heavy will work.

To keep the payload from bouncing after it hits, or to hold it at a specific height above a road, sharp stakes can be lashed to either end that will stick in soft dirt. Some kind of anchor, such as a tree, fence post, etc., set behind either end will strengthen the block.

A payload that has to hit a windshield to be effective

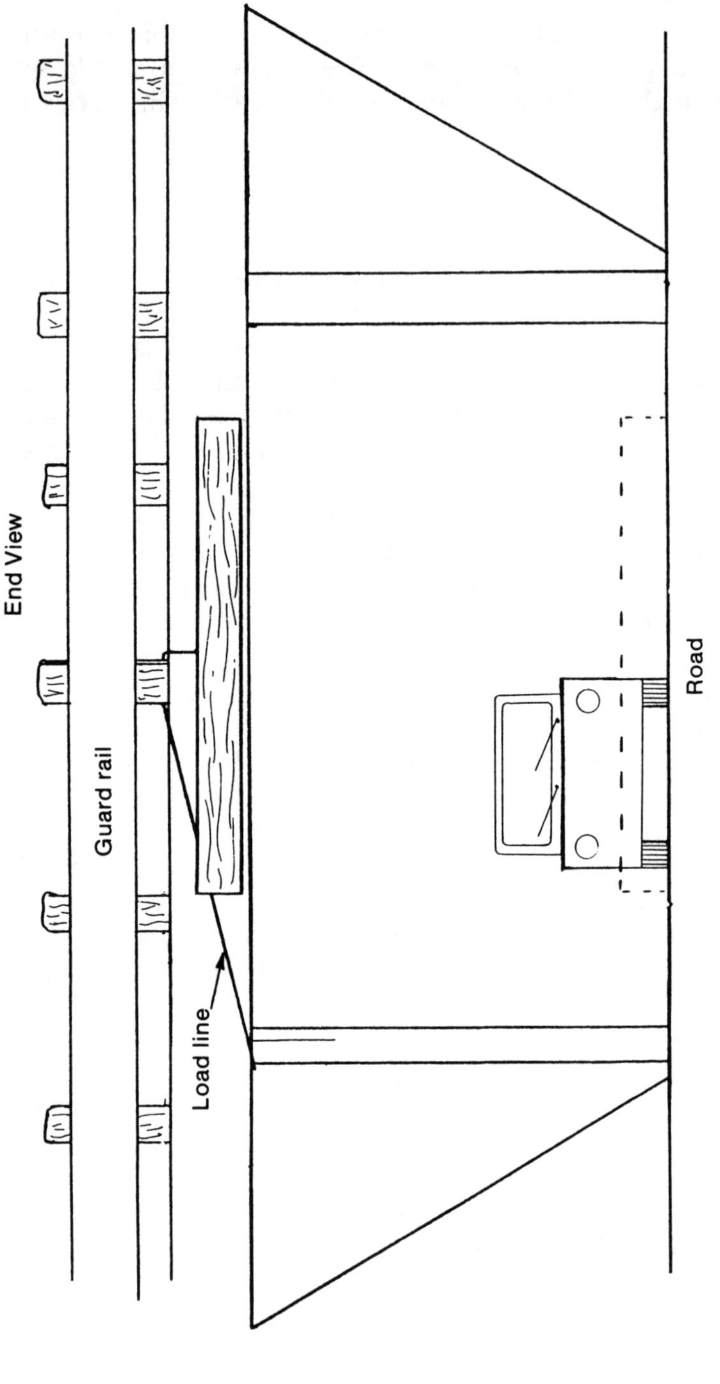

Fig. 24. A payload can be hidden underneath an overpass by placing a board between support beams, as shown. A moveable scaffold can be made in the same way, using two boards.

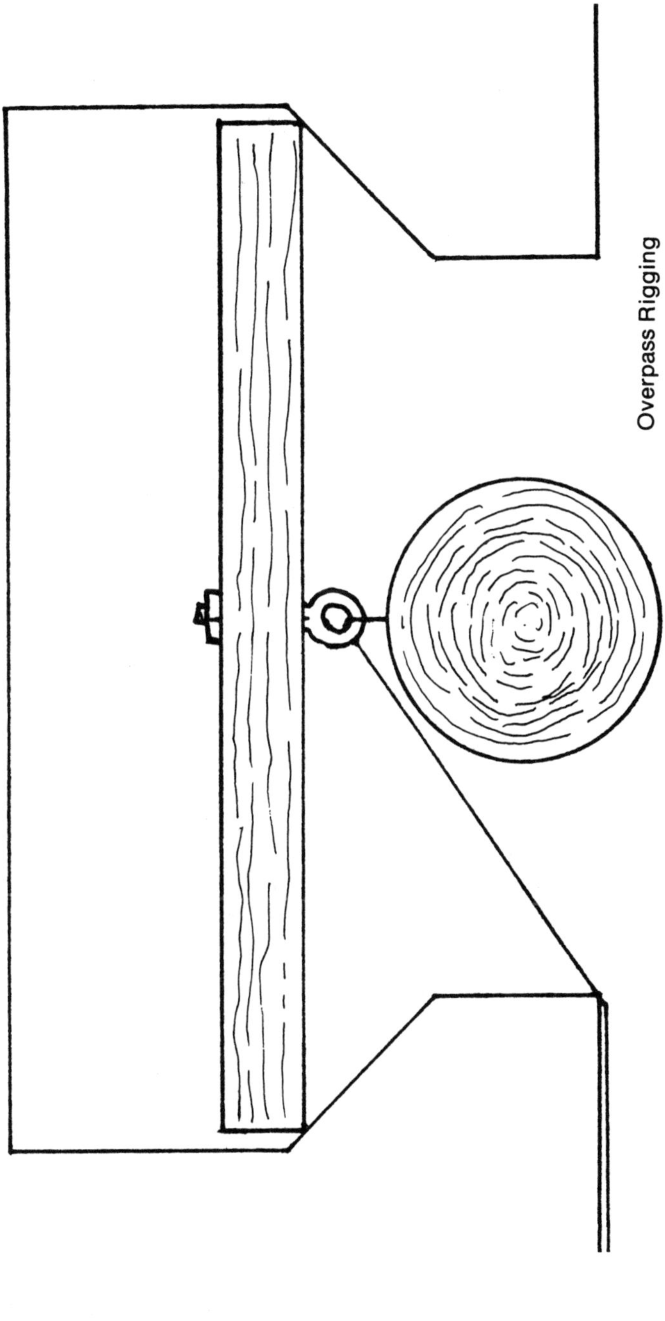

Overpass Rigging

(i.e. manhole cover) can have a knot or other obstruction tied in the load line that will not pass through the pulley, to hold it suspended at the correct height. Obviously this requires a very strong overhead anchor and load line. A long payload such as a railroad rail can drop into brackets at either side of the road that will hold it suspended.

Miscellaneous Weight

Enough weight in any form can kill or injure a person, or wreck a vehicle. Many objects can be used very effectively as is, or with minor alterations.

Hard, heavy objects such as barbell weights, bricks, building blocks, boulders, steel pots, iron skillets, household appliances, and vehicle and machine parts, can all work very well. They can all be easy to attach load lines to, because they do not have to hang in any certain way. A clove hitch, noose, or sling can be used to attach a load line to almost anything.

Soft, heavy objects like bags of flour, sand bags, dog food, and so on can do substantial damage if they fall far enough.

Vessels such as cooking pots can be filled with weight and then taped or glued shut.

Practice Payload

As with all martial arts, skill at boobytrapping comes from practice. Cardboard boxes, pillows, folded or rolled blankets, etc. can be used to simulate real payloads. Water can be dumped from vessels to simulate fuel. Fireworks and blanks can simulate real munitions.

A dummy payload can be used to test a mechanism before it is armed with a real payload. A trap with a dummy payload can be set on friendly territory to hit friendly forces. This will give good indication of how well a certain mechanism, location, or concealment, will work against an invader. It will also make the friendly forces well aware of the trap, so that when it is armed with a real payload they do not fall into it themselves.

Firing Pin Driver

A fired primer is a noise-maker, and so can startle an intruder or be an alarm. It can initiate fire or explosion.

A primer not set in a shell case can be fired simply by being set on a hard surface and hit. A primer in a shell case requires a firing pin.

To fire a primer set in a cartridge, the cartridge must be held securely in some way. Since the weight will fall straight down, the cartridge must stand straight. It can be taped or glued to a wall, a nail can be bent around it; or a rimmed shell can be set in a hole in a board or tin can.

The firing pin can be a nail, screw, any pointed bullet, or anything else that has a suitable shape and hardness. The end that contacts the primer should be as much like a factory firing pin as possible. Nail points should be rounded off.

The firing pin must be centered over the primer, and be straight up and down. It must be able to travel down far enough to fire the primer. Since it is in contact with the primer to begin with, that distance is very small.

It can be held in place with any mouldable or compressable material, such as clay, foam rubber, chewing gum, or styrofoam. Things like wood, soap, and the stalks of some plants can be cut to the right shape and a hole drilled or punched in them for a firing pin. In any case, the firing pin and whatever holds it must be secured to the cartridge so that it remains centered over the primer. This can be done with tape, glue, or in some other way.

A very easy way to fire a primer set in a cartridge is with two pair of vise grips and a phillips head screw-driver.

The cartridge is locked in the jaws of one pair of vise grips, and held suspended primer side up. The vise grips

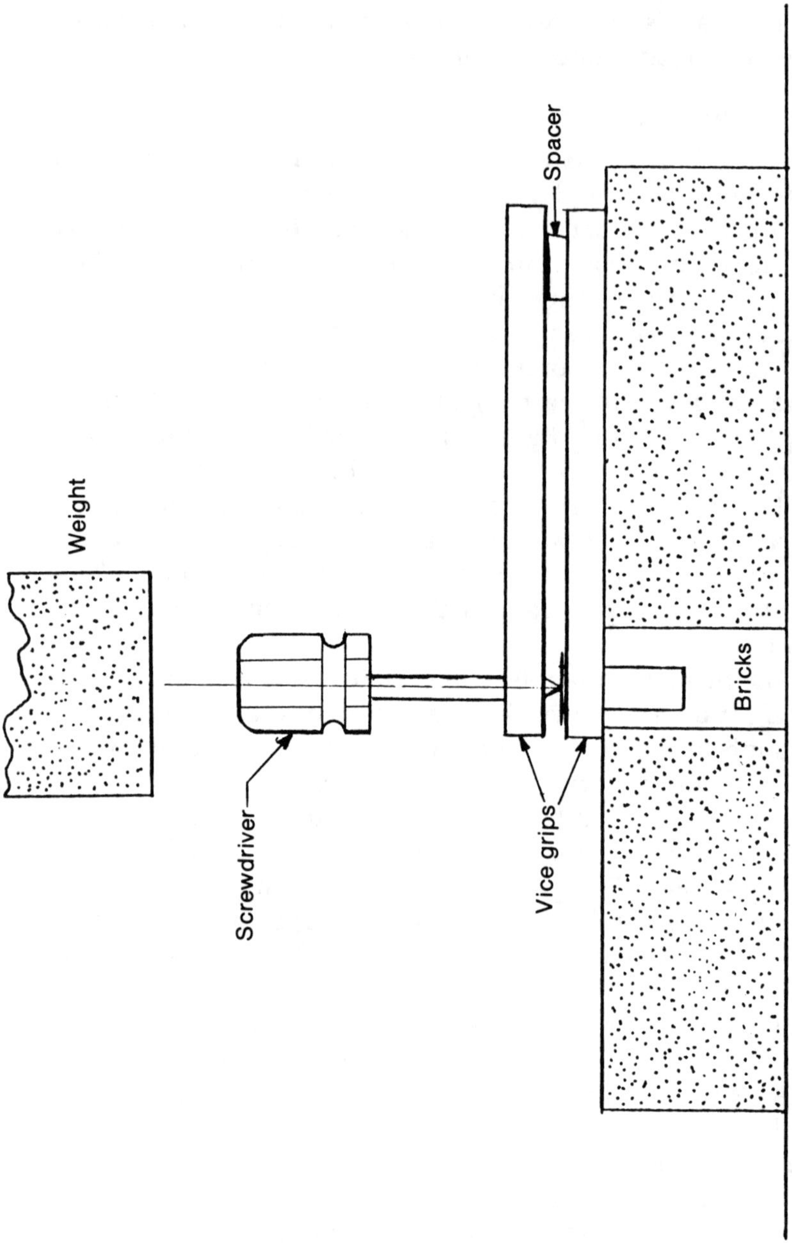

Fig. 25. Noisemaker. Vice grips hold a cartridge and a phillips head screw-driver. A falling brick drives the screw-driver into the primer.

lay between supports, one on either side of the shell. These supports can be two sides of a building block, two bricks, or two boards. They should both be the same height.

The screw-driver (or any other firing pin) is held by the second pair of vise grips, and centered over the primer. It protrudes about an eighth to a quarter inch below the jaws. In order to be level, the handle end of the vise grips sits on a spacer equal in width to the first pair. If no spacer is available, the second pair can lay on top of the first pair.

A brick is hung a foot or so above the screw driver handle. When it drops, the primer fires.

An easier way to make a noise-maker, if primers not set in a cartridge are available, is to set one or more striker side on a flat surface, and drop a brick on it. Rigging should be done so as to cause the brick and surface to come together flat against each other.

Electrical Switch

This can be used to initiate an explosion or fire, activate an alarm (i.e. turn on a loud radio), and a variety of other things. Setting one or more to turn on lights when doors and windows are disturbed can convince a potential intruder that someone is home.

The easiest way to make a switch is to have the bare ends of two leads come together (Fig. 26). This can be done via a tripline alone. A dropping weight, however, will be more certain of causing a good connection, and will maintain the connection even after pressure has been taken off the tripline.

One lead is attached securely to an anchor, and the other is attached to a weight, so that when the weight falls the two are pulled together (Fig. 27). This can usually be done with a very simple balance trigger.

A dropping weight can also turn on or off a conventional switch.

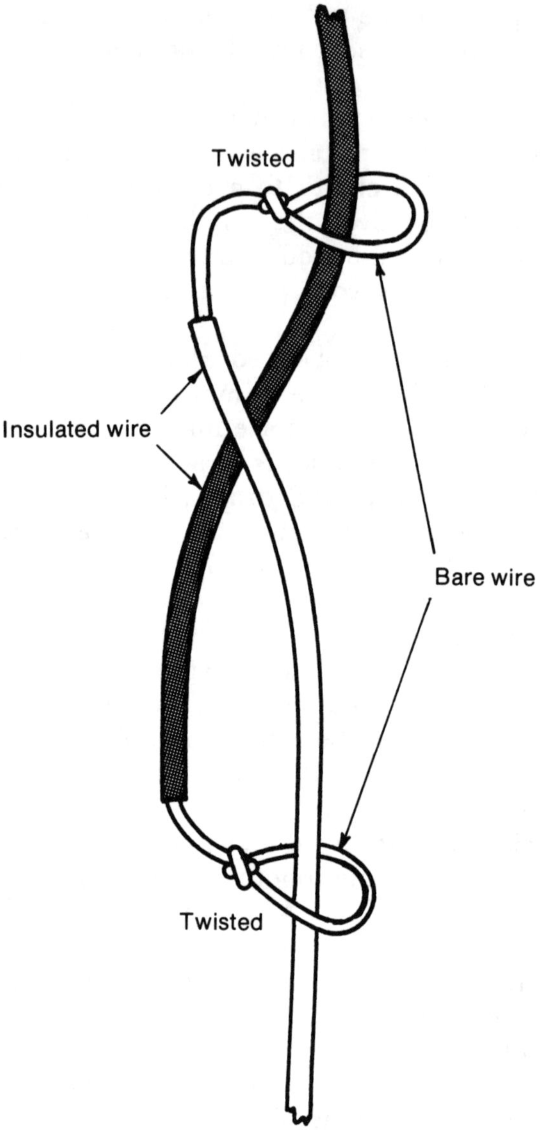

Fig. 26. Electrical switch. When the two leads are pulled apart, the bare ends come together.

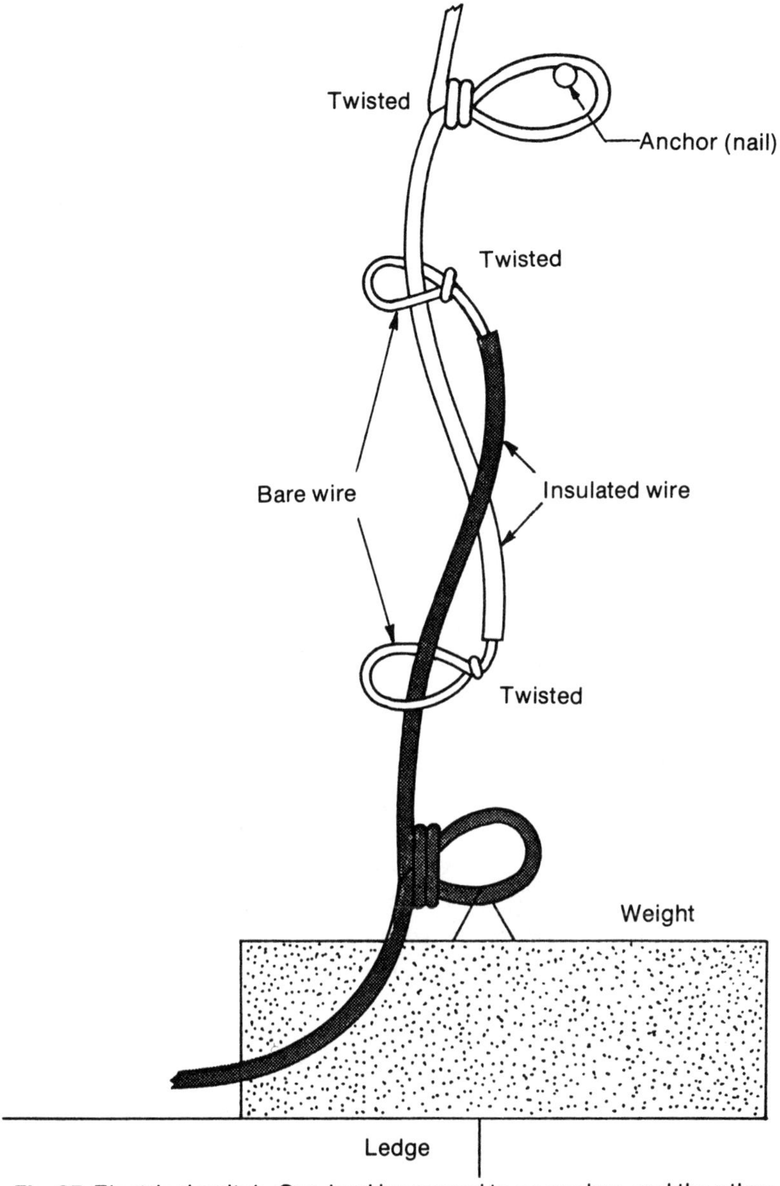

Fig. 27. Electrical switch. One lead is secured to an anchor, and the other is attached to a weight. When the weight drops, the bare ends are pulled together.

Trigger Puller

Any firearm can be set to fire when a boobytrap trigger is activated. This can be done quickly and easily, and can be very effective. A person alone in a house can guard multiple entrances if he has multiple guns. An attacker can be led to believe that a house has multiple occupants, or that an unoccupied area is occupied. A person who has to leave guns behind in an area that will be occupied by an enemy can rig the guns to inflict casualties.

There are many ways that force can be transferred from a tripline to a trigger. One is via a falling weight. The tripline causes the weight to fall, and the weight pulls the trigger. This will often be the easiest and most dependable way for a gun to be fired. Since the weight will maintain pressure on the trigger even after pressure is taken off the tripline, a fully automatic gun will continue to fire until it runs out of ammunition.

The amount of weight required to pull a trigger varies from gun to gun, and also depends on how the rigging is done. If a gun is set to fire full auto, the weight should be substantial enough to pull the trigger without falling. Otherwise, less weight coupled with falling can produce enough force to pull the trigger.

The force must be channelled so as to pull straight back against the trigger. The gun must be secured so that the force moves the trigger instead of the whole gun.

There are several ways this can be done with any gun. One example will be given in detail.

An AR15 rifle is secured to a flat surface (i.e. table top) with lashing, weights, or in some other way. If it is capable of full auto, the selector is placed in that position. Since an AR15 pulls toward the top of the gun when fired, it is laid on one side, and pointed so that the pull will move it along a pre-determined course.

It is secured at only one point. The flat surface and one point attachment allow it to move along one plane but remain fixed on another. Position of the fixed (pivotal) point on the gun will effect the amount of force required to move the gun, and so how fast it moves as it fires. This in turn determines how the bullets are dispersed. Weights can be set against the gun to oppose its travel, and so slow it down.

Weights to pull the trigger can be bricks, or dirt contained in something like a shirt.

Rigging a line to pull straight back on the trigger is very easy with an AR15, because of the rear sling attachment. The line runs from the trigger through it and is attached to the weight. The line is made with the rifle's sling, a boot lace, or any other rigging material.

The weight can be triggered to fall the same as any other deadfall. One easy way is to set the weight on the gun stock (balance trigger). It can be activated by a tripline, or be so balanced as to be disturbance activated.

Grenade Activator

A grenade can be set to drop in the same way as any other weight. The pin can be pulled by a line leading to an overhead anchor. Another line can be used to stop and hold the grenade at a certain height. The grenade can be explosive, gas, or incendiary. An incendiary grenade makes an excellent fuel igniter.

A grenade pin is a cotter key, which holds a spring loaded lever in place. The grenade should be hung so as to allow the cotter key to be approximately straight up and down, so that it will pull straight out. If the pin is split, the ends should be as close together as possible. The pin should be as straight as possible.

There should be nothing to keep the lever (handle) from springing free of the grenade, such as a load line tied around both it and the grenade.

Match Head Igniter 1

A simple igniter can be made using one or more books of matches, and tape. It can be used to startle an intruder, as an alarm, to light a fuze, or to initiate fire. Match head igniters can be made in a few different ways. The one discussed below is easy and fast. If the instructions are followed closely and construction done carefully (i.e. folds and tears made along straight lines), and if new matches in good condition are used, it can be very reliable.

The book is carefully removed from the matches by pulling it loose from the staple, and flattened out. A strip of cardboard is torn or cut off the bottom, leaving a strip equal in width to the striker. That strip is folded up behind the striker. The book is then folded in half lengthwise, with the striker side out. The folded up strip causes the striker to bulge outward slightly. Fig. 28

The matches are folded in half over the striker, with the bottom of the heads at the top of the striker, and taped in place. It is important that match heads surround the striker completely, and that no tape touches the striker. Tape is wrapped tightly around the matches so that there will be pressure between them and the striker when the striker is pulled out. The wrapping should not be so tight as to make the striker overly difficult to remove. The tape extends over the match heads themselves, so that they squeeze in over the striker. The ends of the match heads must be free of tape, so that the flame has a place to go. Otherwise the device will produce more smoke than fire. Fig. 29

Moisture of any kind on the match heads or striker can render the device inoperative. This includes liquid fuel, which should not ever contact the igniter prior to ignition. Hands should be kept free of sweat during construction.

Ignition occurs instantly after the striker is pulled out. Burn time is approximately one second, with a small flame continuing after the match heads have burned out.

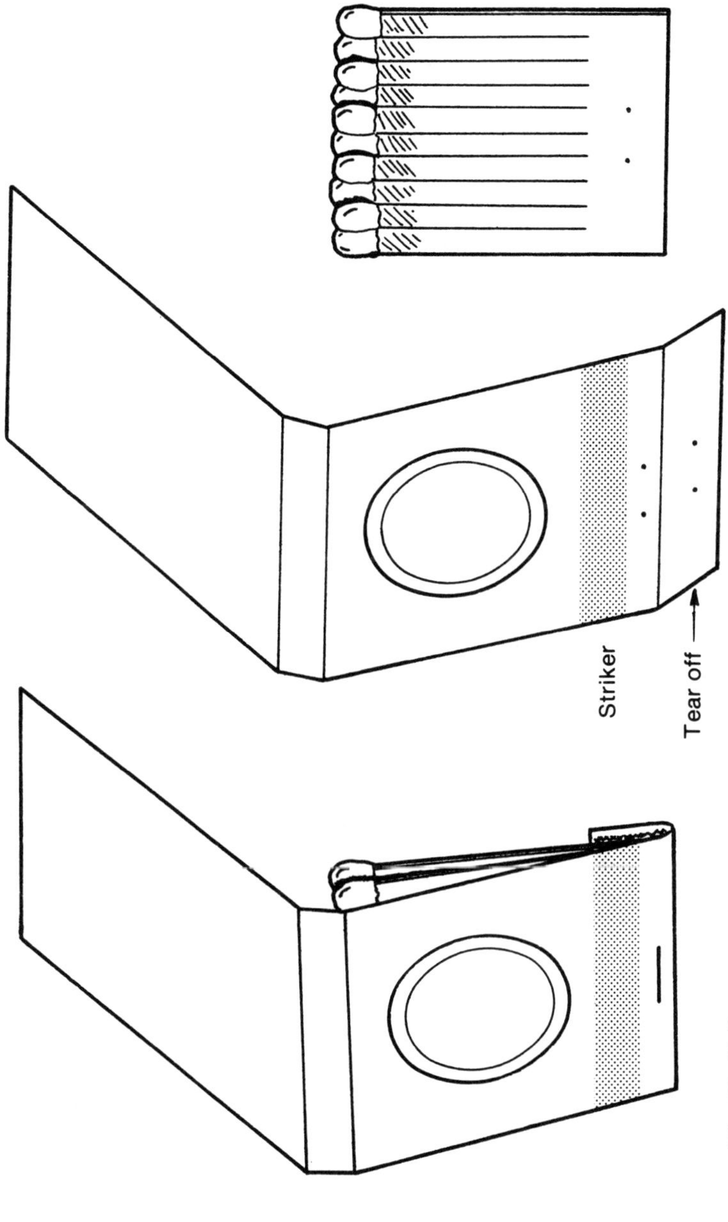

Fig. 28. Striker assembly.

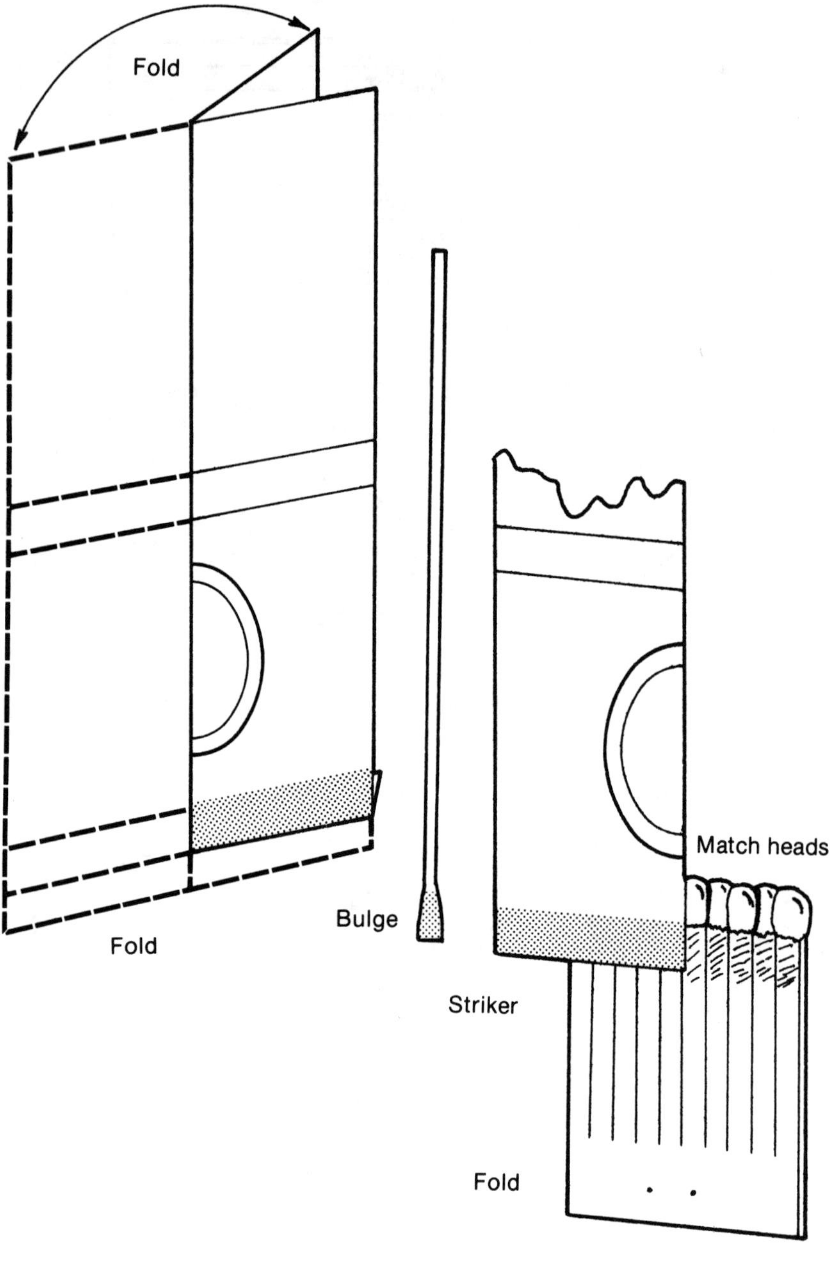
Fold
Fold
Bulge
Match heads
Striker
Fold

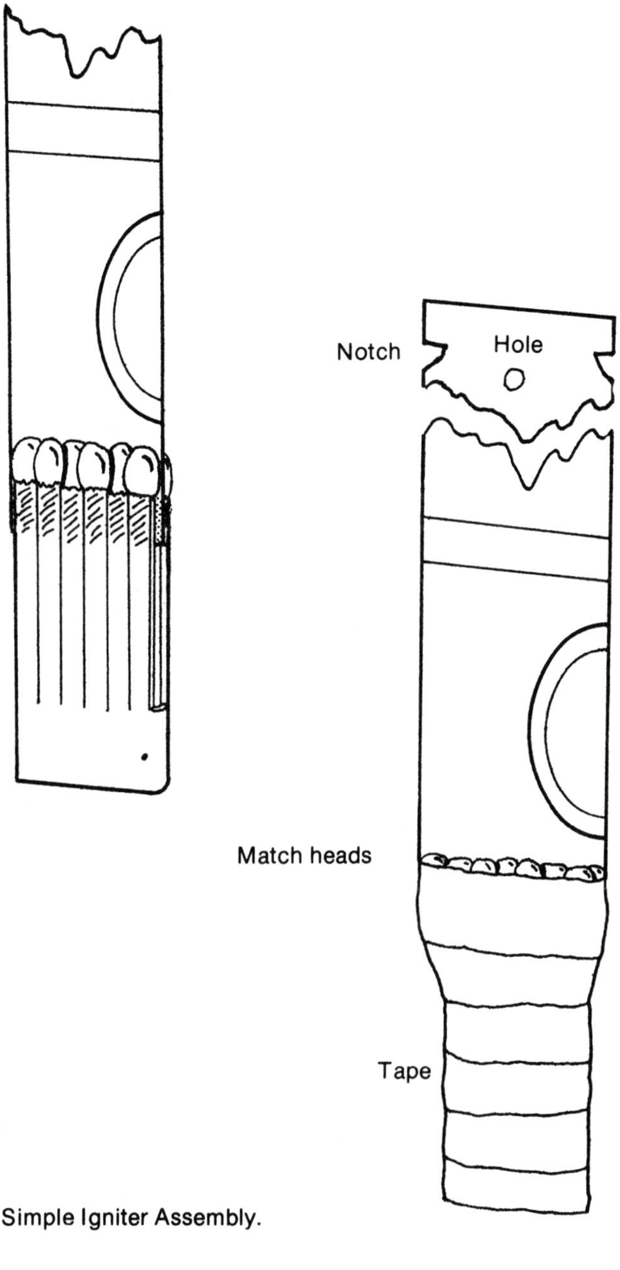

Fig. 29. Simple Igniter Assembly.

Power can be increased by taping more matches around the original ones. If matches are attached with all of the heads touching, size and intensity of the flame will be increased, but burn time will remain about the same. Adding tape or distance between heads will increase burn time.

This igniter can also be made to blow fire out the bottom (opposite the direction the striker is pulled) instead of the top. It is made the same as the one described above, except that the striker assembly is turned over, so that the striker is above rather than below the match heads, and pulls down between the match sticks. The match sticks should not be taped so tight as to make removal of the striker difficult. Only the match heads need to grip tight. Fig. 30

A safety can be installed on any match head igniter. This is done by folding the striker down and taping it to the matches, so that it can not be pulled out. The end of the tape is folded over itself to form a tab. This allows it to be removed easily.

Match Head Igniter 2

Match heads can be used more efficiently if cut loose from their cardboard sticks. An igniter made with match heads alone will have more power for its size, be more reliable, and can do more things. It can be made to blow fire out the top, sides, or bottom.

It is made with a quantity of safety matches in good condition, tape and/or glue, and a tool to cut the match heads off with. This is best done with wire cutters of the type that pinch rather than scissor. A sharp knife can be used, but is much slower. Some type of wadding material, such as tissue paper, or cloth is very useful but not essential.

The match heads are contained in some sort of vessel. Its size determines the amount of match heads, and so the

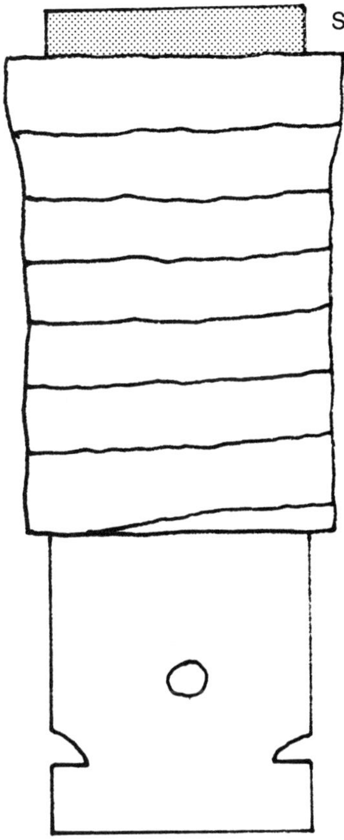

Fig. 30. Inverted igniter. This igniter will blow fire opposite the direction the striker is pulled.

amount of fire the device will produce. It is approximately cylindrical, with the top left open. Holes in the vessel will determine which way fire blows. To make an igniter that will blow fire out the sides or bottom, holes are punched in the sides or bottom. They must not be so large as to allow match heads to fit through them. If no holes are put in the vessel, fire will blow out the top.

The best material to make the vessel out of is light cardboard. The box that match books come in works as well as anything. Match books themselves, with the matches removed, can be taped or glued together to form a suitable vessel. Many other things, such as aluminum cans, can be used. But if there is any reason to hide the fact that a fire was started with the device, all components should be flammable.

The exact shape of the vessel depends to a large extent on what it is going to be atached to. Since the striker will always be pulled out the top, it is usually good to make the bottom larger than the top. That will make the igniter easy to attach to things.

It might seem as though something like an emery board or fine sandpaper would make a good striker, but there really is no better striker than the one that comes from the factory. The striker assembly is the same as described for match head igniter 1. The two halves are glued or taped together in this case, to keep match heads from getting between them. If taped, all tape must be kept on the inside.

The striker is held in the center of the vessel, extending to the bottom, as the vessel is filled with match heads. The vessel should be shaken occasionally as it is being filled, to help the match heads settle in place. They should be packed down tightly.

Wadding is stuffed in around the striker. It must hold the match heads in place before and after ignition, and allow the striker to be pulled through uninhibited. Something soft, such as tissue paper, works well. If nothing else is available, pieces cut from match books and tape can be used. If glue is available, the wadding is glued to the vessel, and partly saturated with glue. It is very important that the striker not be glued to the wadding. If no glue is available, the vessel sides are folded down over the wadding, and taped down securely. Fig. 31

The line that will activate the striker should attach to the center of the striker, and the pull should be as straight as possible. It can be attached by a loop fitting into small notches in either side, or through a hole in the center. It should then be reenforced with glue or tape.

Burn time is approximately one second. More match heads will not significantly affect burn time, only size and intensity of flame. Adding slow burning materials (i.e. tape, paper) between layers of match heads can extend burn time.

Molotov Cocktail

This is the traditional people's fire bomb. It consists of a breakable vessel, fuel, usually a wick, and fire. It can succesfully attack individuals, groups, and vehicles, including armored vehicles. It can also be used to ignite more fuel. It can be used in combination with other mechanisms that do nothing but spill fuel. Something like an old barn made of dry wood can be turned into an inferno in a short amount of time with liquid fuel and fire applied to it correctly. Fire should originate from escape routes.

Used together with one of the igniters just described, a molotov cocktail can be set and triggered the same as any other dropping weight.

The vessel is anything that can contain liquid fuel and break. Large vessels will obviously result in more power, because they can contain more fuel, fumes, etc. Fuel can be gasoline, alcohol, cleaning fluid, and so on. If possible it should be thickened with oil or soap to extend burn time

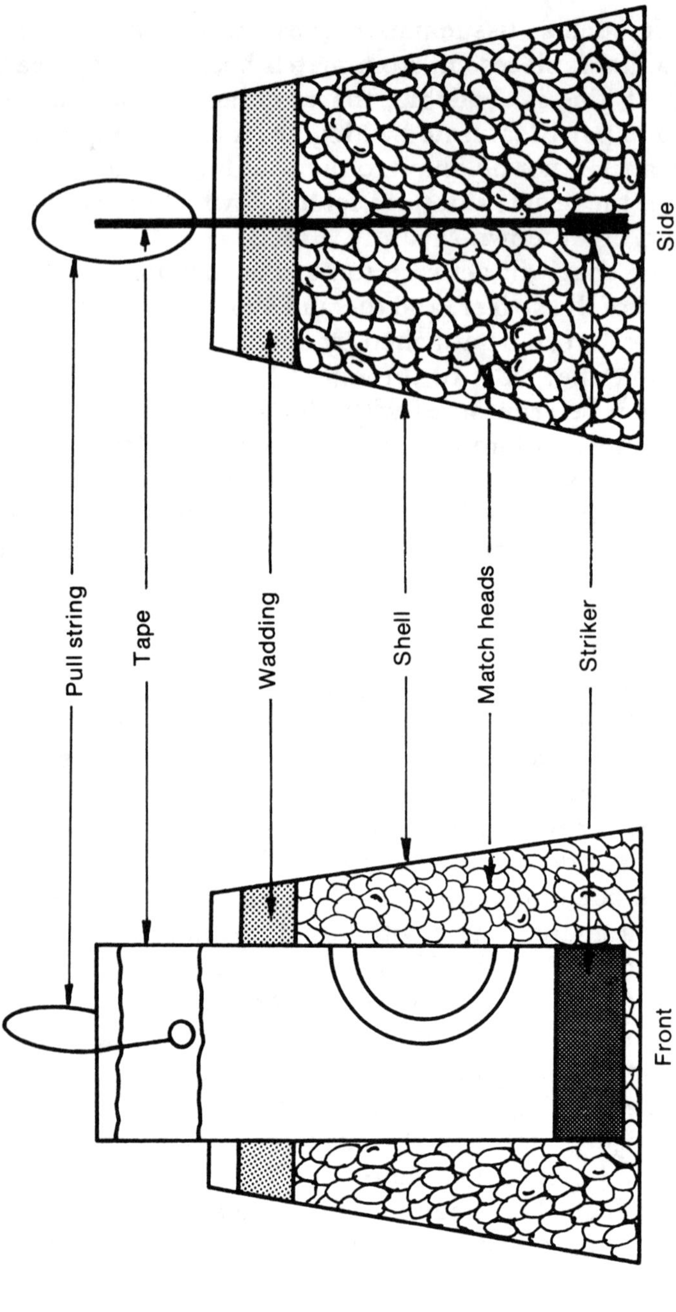

Fig. 31. Match head igniter II. Construction detail.

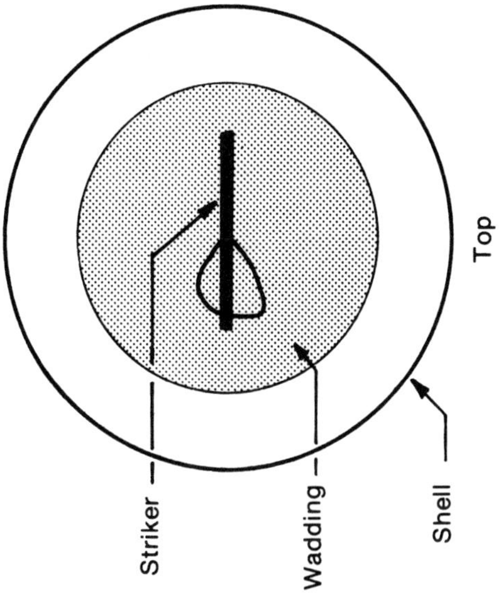
Striker
Wadding
Shell
Top

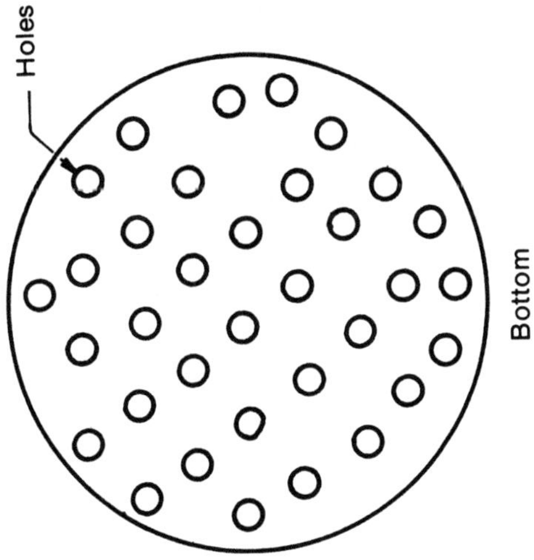
Holes
Bottom

and allow it to stick to things.

Either Igniter 1 or 2 can be glued or taped to the vessel. A line runs from the striker to an anchor, with enough slack to cause the weight of the falling fuel vessel to jerk it out. If ignition occurs just prior to the vessel breaking, the fuel will ignite, and should spread.

To be more certain that fire reaches the fuel, and so simplify timing, a wick can be placed between the fuel and igniter. This can be any absorbent cloth, such as cotton. One end is placed in the fuel, and the other in the area that the igniter will blow fire to. It must be far enough away from the igniter so that it does not contaminate it with liquid. A wick should suck fuel out of the vessel and stay moist, but may not. To be certain that it has fuel on it when the trap is triggered, it can be coated with burnable oil (i.e. vegetable oil, bacon grease, motor oil, etc.) which will not dry out.

Use of a wick will allow fuel to slowly evaporate, because the vessel must remain open to accomodate it. In most cases a wick will not be necessary.

A better way to extend burn time is to attach more matches to the outside of the vessel. Providing they are not in close contact with the ignition matches, burn time will be significantly extended, even up to several seconds.

Fire can be initiated in other ways as well. A chemical igniter, (see **Improvised Munitions Handbook**), is well suited to this. A rag (wick) can be stuffed into the bottle and lit. A flame will burn at the mouth of the bottle until the bottle breaks, at which time it will ignite the fuel. Since fuel is consumed as the flame burns, the more time there is between setting and triggering, the less power the trap will have. Burn rate is determined by the amount of wick sticking out. The flame should be substantial enough to not be blown out when the vessel falls.

An igniter that blows fire out the bottom can be set in the top of a jar to blow fire directly into the fuel as the vessel falls. It is mandatory that the fuel receive sufficient oxygen to burn. Fuel in something like a pop bottle will not be ignited by an igniter stuffed into its mouth, because the igniter will use all the available oxygen, and leave none for the fuel. The container must have a large enough mouth, and there must be sufficient open space for air to reach the fuel. A food jar works very well. One large hole is bored in the center of the lid to receive the igniter, and small holes are punched around it for air.

Holes in the lid simplify rigging by acting as brackets for the line to be tied through. the jar should hang perfectly level, to keep fuel from reaching the igniter and contaminating it.

During transport, a lid without holes should be on the jar. The lid with the igniter is installed just prior to setting the trap.

No matter how the fire is initiated, the vessel must be broken in order for it to spread. This can be done by dropping it onto a hard surface, such as a rock or pavement. Breaking can be at any level, including above the target, so that fire rains down on it. If a moving vehicle is targeted, the vessel need only sit in the vehicle's path. This is done by attaching a line from it to an overhead anchor that will hold it suspended at the desired height (windshield level). This line must have more slack than the striker line. The two lines can be one and the same, or a knot in the load line can hold it. In any case, the line must be attached so as to not break the vessel when it catches it. A shock absorber, such as a large rubber band, might be used, providing it does not have enough strength to make the vessel bounce, which could cause it to be missed.

If necessary, any of the above devices can be made more or less waterproof by covering them with an inverted palstic bag, that will remain fixed and allow the payload to fall free on triggering.

Dumpers

A vessel can be hung and set so as to turn over and spill its contents on triggering. The vessel can be a can, bucket, or drum. Its contents can be fuel, paint, acid or anything caustic. One or more dumpers can mix chemicals together. Things like snakes and spiders can be contained in a vessel and dumped on a target. The mechanism is more or less the same in each case.

A load line attaches to the top of the vessel and holds it suspended upright. Another line is attached from the bottom of the vessel to an anchor, so that when the line releases, it will cause the vessel to turn over and hang upside down.

The vessel will tip directly opposite the point where the bottom line is attached, and the contents will spill in that direction. The payload is aimed by setting it to tip toward the target area. Fig. 32

Fuel

Liquid fuel will be gasoline, alcohol, kerosene, cleaning solvents or varnish. It can be made more effective by being thickened with oil.

When using fire as a payload it should be remembered that smoke is a much greater killer than flames. Smoke from such things as fur, wool, silk, and plastics (from furniture, rugs, clothing, curtains, etc.) is among the most lethal. Since smoke rises and fire burns upward, fire started in lower parts of a structure will effectively attack people above it. Fire started in a basement will quickly send smoke and flames to all floors above it.

For more information on fire starting, consult books on safety, fire prevention and camp craft.

In most cases the fuel will be ignited just before or as it dumps, so as to be burning when it contacts the target. The dumper and igniter can be part of the same system. Or they can be separate, individually triggered mecha-

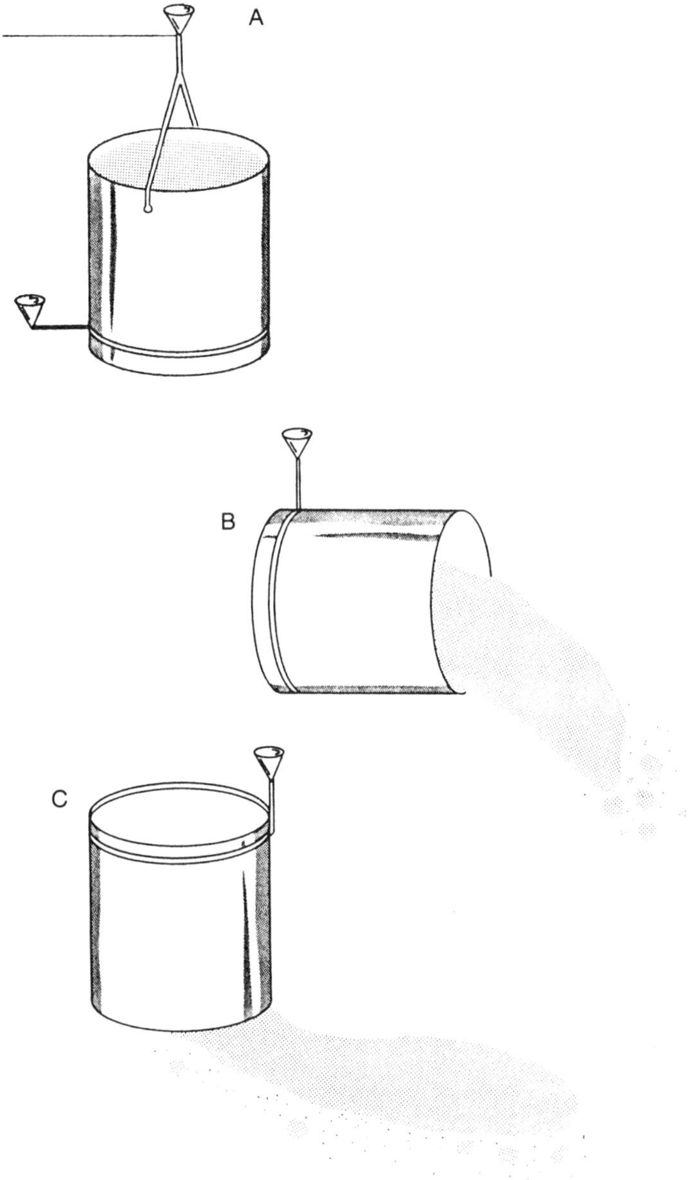

Fig. 32. Dumper aim. The vessel tips opposite the bottom anchor, and spills its contents in that direction.

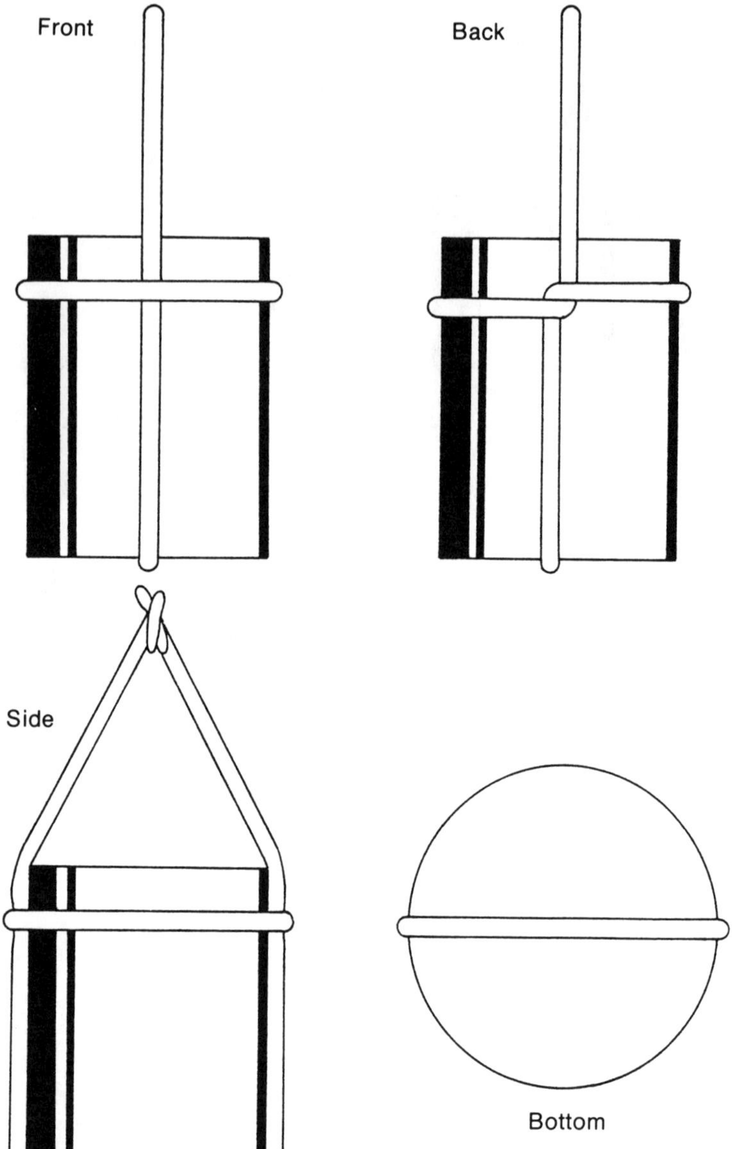

Fig. 33. Dumper rigging. A dumper can be hung either by tieing a line through holes in the top, or by tieing a line around the vessel as illustrated. The line on the bottom must be well centered.

nisms. For example, fuel is dumped on the target as a warning (i.e. "Do not go any further into this area or you will be torched."), and as proof that the threat is real. Igniters (fireballs, etc.) are small and are easy to set, hide, and hit targets with, especially since fire needs only to reach fumes to ignite the fuel. They can be placed to burn the target if the warning is ignored. Fuel used in this way should be oil based, so that it can not be dried off easily.

Multiple dumpers used in combination can quickly turn an entire structure into a torch. Only one needs to have a trigger and igniter attached to it, with the rest triggered by fire. But multiple igniters will both increase dependability (because one might fail), and spread fire faster. Fire should originate from escape routes, and the initiating dumper should be triggered from deep inside the structure.

There are several ways for an igniter and dumper to be incorporated. It is important to note that low grade fuel dumped directly on a weak igniter might extinguish the igniter and not burn. Liquid fuel dumped directly on an igniter before ignition is very likely to render the igniter inoperative, unless it has been made moisture-proof. So the most dependable way is to blow fire into the fuel, then dump the burning fuel.

This is done by attaching an igniter that blows downward to the top inside of the vessel. A line leads from the striker to an overhead anchor, with just enough slack to cause the striker to be jerked out by the force of the falling vessel. The line attached to the bottom of the vessel has slightly more slack than the striker line, so that the vessel tips an instant after the fuel is ignited. Fig. 34 illustrates the correct timing sequence. Note that one line pulls the striker and dumps the vessel both. This simplifies rigging, and conserves fuel.

A simple igniter that blows fire out the top can be set to ignite fuel as it dumps. This works well with small dum-

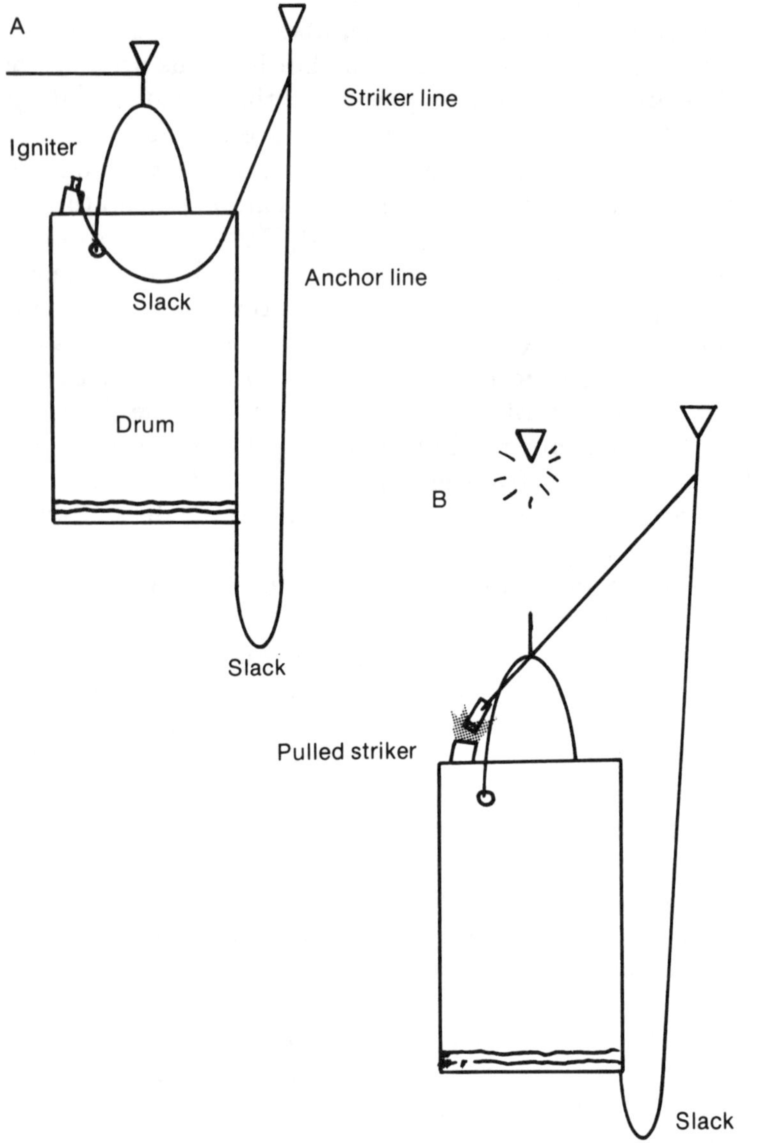

Fig. 34. Dumper/igniter timing sequence. A) The striker and anchor line both have slack; the anchor line has more slack. B) When the vessel falls the striker is pulled, and the fire blows down into the fuel. C) The anchor line grabs and tips the vessel, spilling burning fuel.

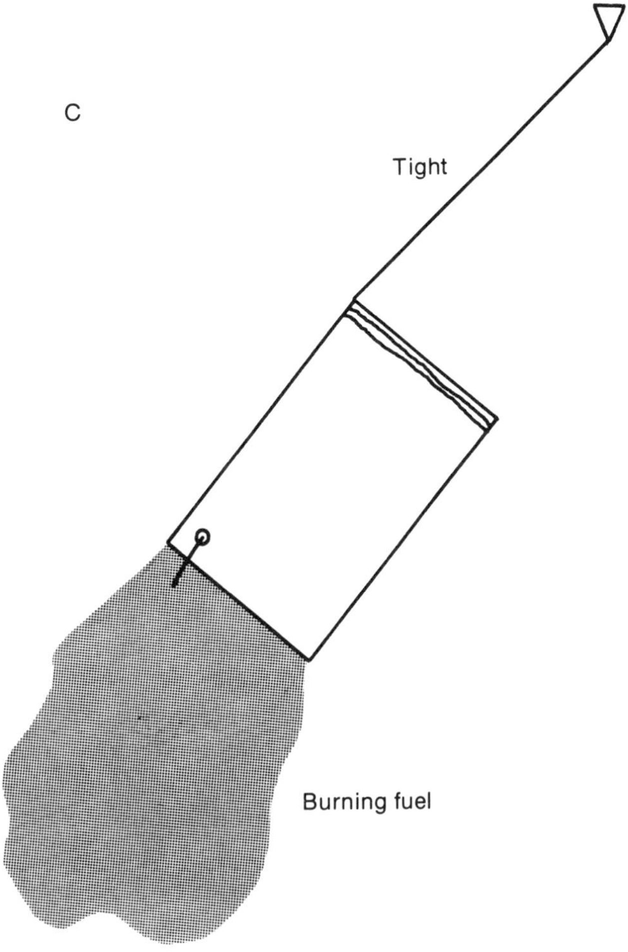
C
Tight
Burning fuel

pers, such as drink glasses. It is attached to the outside of the vessel just below the top lip, so that its fire will reach the fuel as it dumps, but liquid will not pour directly on it. It must be attached opposite the line that will dump the vessel, so that the vessel will tip toward it. A piece of absorbent cloth can be laid over the lip with one end in the fuel and the other where fire will reach it, to act as a wick. The end must be well enough away from the igniter to keep from contaminating it with liquid. One or more extra books of matches can be added to increase burn time. Fig. 35

An alternate method is to hang the vessel with a burnable line that runs through holes in the top and forms a circle. The load line that leads to the trigger is attached to an independent weight (i.e. rock) that pulls a striker on triggering. When the fuel ignites, it burns through the line that holds the vessel upright, and the vessel tips over. An igniter can be set directly over the line, so that it cuts through it with its flame as it ignites the fuel.

This method can be useful when large amounts of fuel are used, because the primary load line does not need to be strong enough to support the heavy fuel vessel. This can both simplify rigging and concealment.

A simpler way to ignite fuel is to leave something like a candle or lamp burning where the fuel will reach it. Liquid fuel poured directly on a small flame will likely extinguish the flame rather than ignite it. So something between the vessel and flame to deflect the fuel will make ignition more certain. The liquid does not need to ever contact the flame; fumes are what burn. A tent made out of absorbent material built over the candle will keep liquid from reaching the flame. It is important that the tent allow sufficient oxygen to reach the flame. When the fuel hits it, the tent quickly fills with fumes, which easily ignite.

In some cases it will be necessary to cover the vessel, either to keep out rain water, to keep the fuel from evapo-

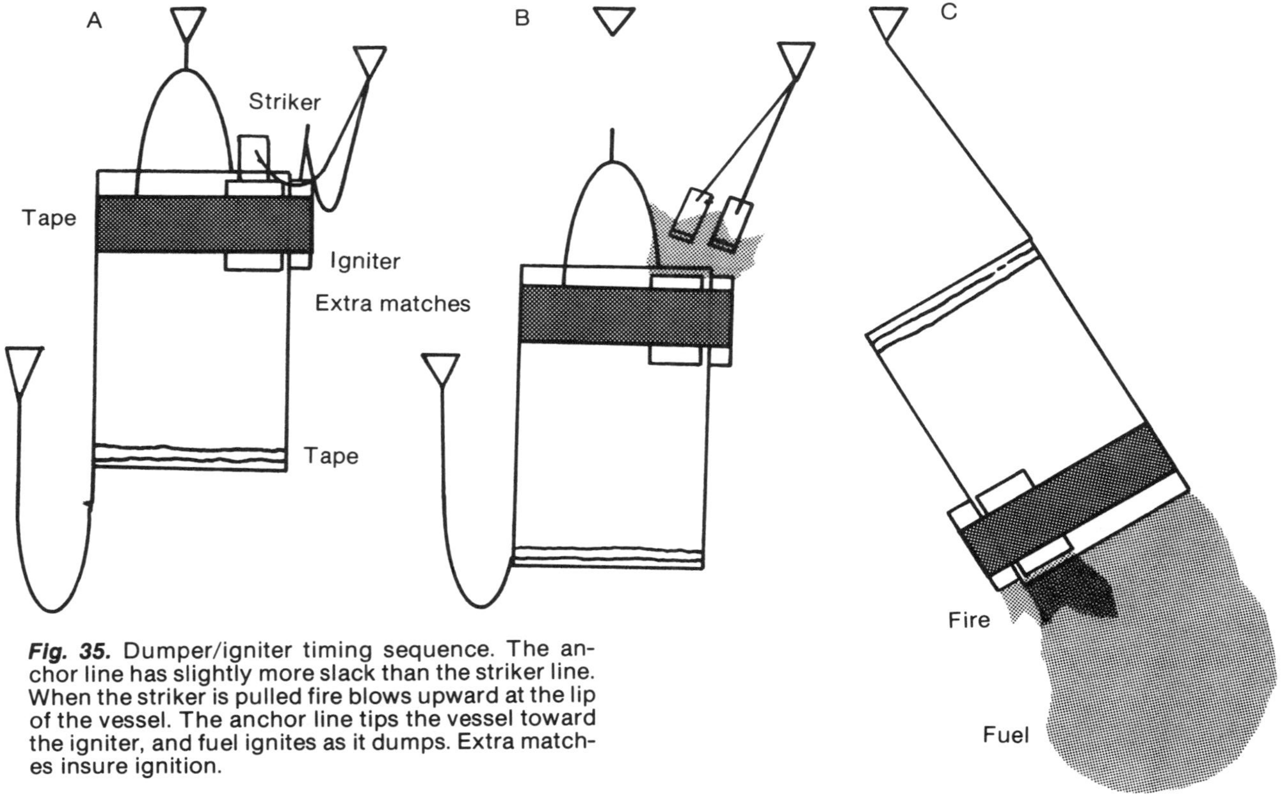

Fig. 35. Dumper/igniter timing sequence. The anchor line has slightly more slack than the striker line. When the striker is pulled fire blows upward at the lip of the vessel. The anchor line tips the vessel toward the igniter, and fuel ignites as it dumps. Extra matches insure ignition.

rating, or to keep the target from smelling fumes. The easiest way to do this is to surround it with an inverted plastic bag (i.e. trash bag). The bottom of the bag is attached to an anchor above the vessel, so that when the vessel drops, it falls free of the bag.

Paint

A vessel filled with paint can be used to mark a target. This can be a good way to discover the identity of a person going into an area he is not supposed to be in, such as a burglar. It can also serve as a warning.

Acids

Strong, concentrated acids dumped on or near a person can produce severe injury or death. Some acids (i.e. sulphuric), will produce fumes when dumped, which can be dangerous, or at least irritating. Nitric acid dumped on organic materials such as paper or wood can produce thick clouds of toxic fumes. It can also initiate fire. Sulphuric acid dumped on a mixture of half potassium chlorate and half sugar will produce fire.

Liquids and Cold

If the temperature is cold enough, any liquid dumped on a person can cause injury or death. The amount of damage done by the liquid depends on the degree of cold, amount of liquid, and time spent in the cold. A person doused with liquid in cold weather must either get into a warm place, or freeze. Less liquid, cold, or time can produce frostbite, which can be very serious.

One problem with this is that if the temperature is low enough to cause being wet to be a problem, it is low enough to freeze water in a bucket. This can be dealt with by using a liquid that freezes at a very low temperature. This includes salt water, alcohol, and antifreeze.

Insect Liberator

Stinging and biting insects, such as bees, wasps, hornets, and flies, can be contained in a breakable vessel that can be dropped and broken on triggering. The vessel must contain air holes. When the vessel breaks, its angry occupants will attack any human in the area.

Power Source

The energy of a falling weight can be channelled to a variety of secondary mechanisms in several ways. Force is transferred from the weight by way of a secondary load line and one or more pulleys, or by a lever system. The same rules of rigging that apply to a primary load line apply as well to a secondary line, the most important of which is: keep the line as straight as possible. Every turn will take some amount of power. Levers will transfer force and also change its nature.

The number of boobytrap devices that can be made in this way is unlimited. A few examples will be given.

Noose

A falling weight can pull a noose tight. Noose traps are especially well suited to small openings such as windows, tunnels, etc., that a target will reach or crawl through. They can be relatively easy to conceal, because the weight does not have to hang directly over the target area. And they can be very powerful, because the weight will be concentrated onto the small surface area of the noose, and the force will be pushing against itself. As with all deadfalls, more weight and a smaller contact surface area (noose material diameter) produce more power.

The best material for the noose is steel wire or cable. Rope, especially nylon, will lose shock force as a result of stretching. But if enough weight is used, any noose will kill or seriously injure and trap the target, providing it does not break. The smaller it is, the better it will cut.

To get maximum effect, the noose should be pulled tight through a small stationary opening (i.e. screw eye), rather than be free running. That way the target will be trapped and squeezed rather than pulled. Two nooses can pull opposite each other.

Figs. 36 and 37 illustrate two ways this type of trap can be used. In each case it is mandatory that the weight fall far enough to pull the noose tight around the target.

If necessary, the distance that the weight has to fall can be cut in half by running both ends of the noose through the stationary opening and attaching them to the weight. That way twice as much line will be pulled through in the same falling distance. If two nooses are used pulling more or less opposite each other, distance that the weight must fall is cut in half again.

A noose can be concealed very easily. In a tunnel it can be pushed into dirt. In a window it can sit where the glass and frame come together, covered with caulking or tape. Small diameter material such as piano wire, can be painted over.

Door Slammer

A slamming door is a good noise-maker, and so can act as an alarm. It can startle an intruder; make him jump for cover. It can convince an intruder that an unoccupied area is occupied. It can also trap him.

If the weight is on the side that the door will pull toward, rigging is easy. For maximum velocity and noise, the line should be attached to the door an inch or so from the hinge side. The line runs from there through a pulley set in the door frame at about the same height. From there the line runs through another pulley, and attaches to a weight. That weight is set and triggered the same as any deadfall. When it falls, the line pulls the door shut. A pretty substantial weight is required to close a door when the line is attached close to the hinge, due to the door's

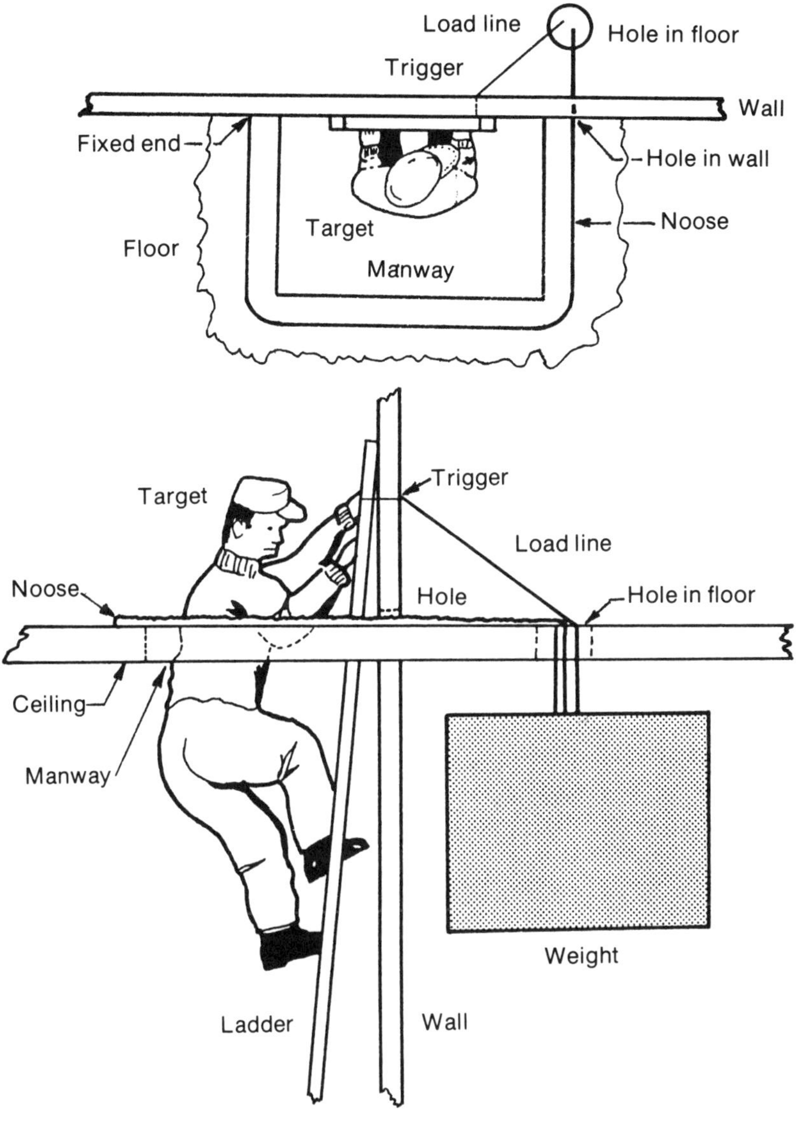

Fig. 36. Manway noose trap. The noose surrounds a manway that leads between floors and is next to a wall. One end is fixed. The other end runs through a hole in the wall, then through a hole in the floor, and is attached to a weight. The primary load line goes through the same hole, and leads to a trigger. The trigger is activated by a ladder rung lever.

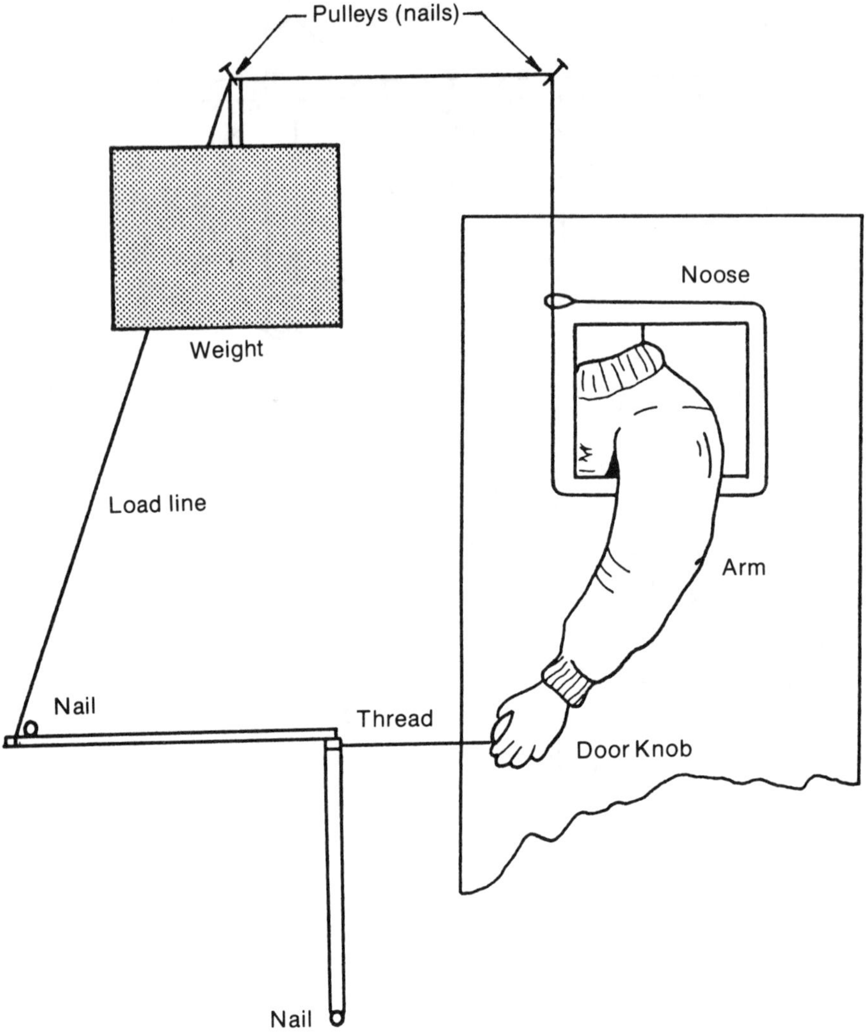

Fig. 37. Arm grabber. Since the small opening will trap the arm, the noose does not need to be run through a stationary opening. The weight and trigger can set as far from the door as is necessary for concealment. A thread leads from the door knob to a hair trigger.

leverage. A heavy weight will slam the door very hard and fast. Fig. 38

The more time the force has to act on the door, the faster it will be moving when it hits. So the door should be open as much as possible when the trap is set.

If the weight and rigging are on the side that the door will move away from, rigging is slightly more complicated. The line attaches to the door as near to the top or bottom as possible, without stopping it from fitting into the frame. The line then runs through a pulley at a corresponding point on the frame. From there the line runs through a pulley, and is attached to a hanging weight.

The same type of mechanism can be used in reverse to pull a door open.

Latch Opener

There are several ways that weights can open various types of latches. Figs. 39 and 40 give two possibilities. With such a mechanism, guard dogs can be kept in a pen until an intruder activates a trigger. The dogs hear the latch open, and push against the door, which opens and lets them out.

Axle Spinner

A weight with a line wrapped around an axle will cause the axle to spin as it falls. The greater the weight and the smaller the diameter of the axle, the faster it will spin.

A silent alarm can be made by attaching one or more flags to an axle, that will draw circles when it spins. The drive wheel of a generator, such as from a bicycle or car, can be spun to produce electricity to fire an electric blasting cap.

Silent Alarm

A falling weight can pull one side of a lever down, mak-

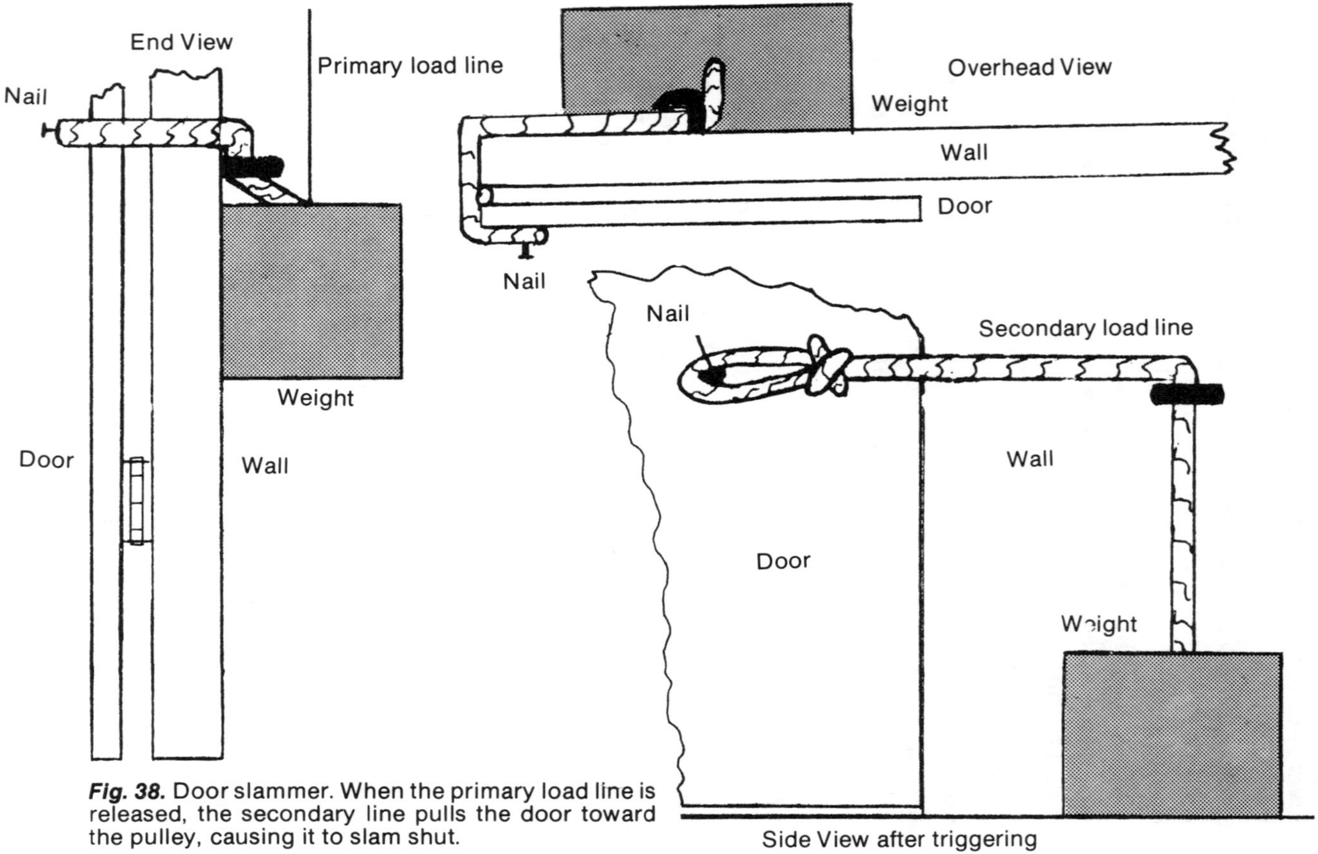

Fig. 38. Door slammer. When the primary load line is released, the secondary line pulls the door toward the pulley, causing it to slam shut.

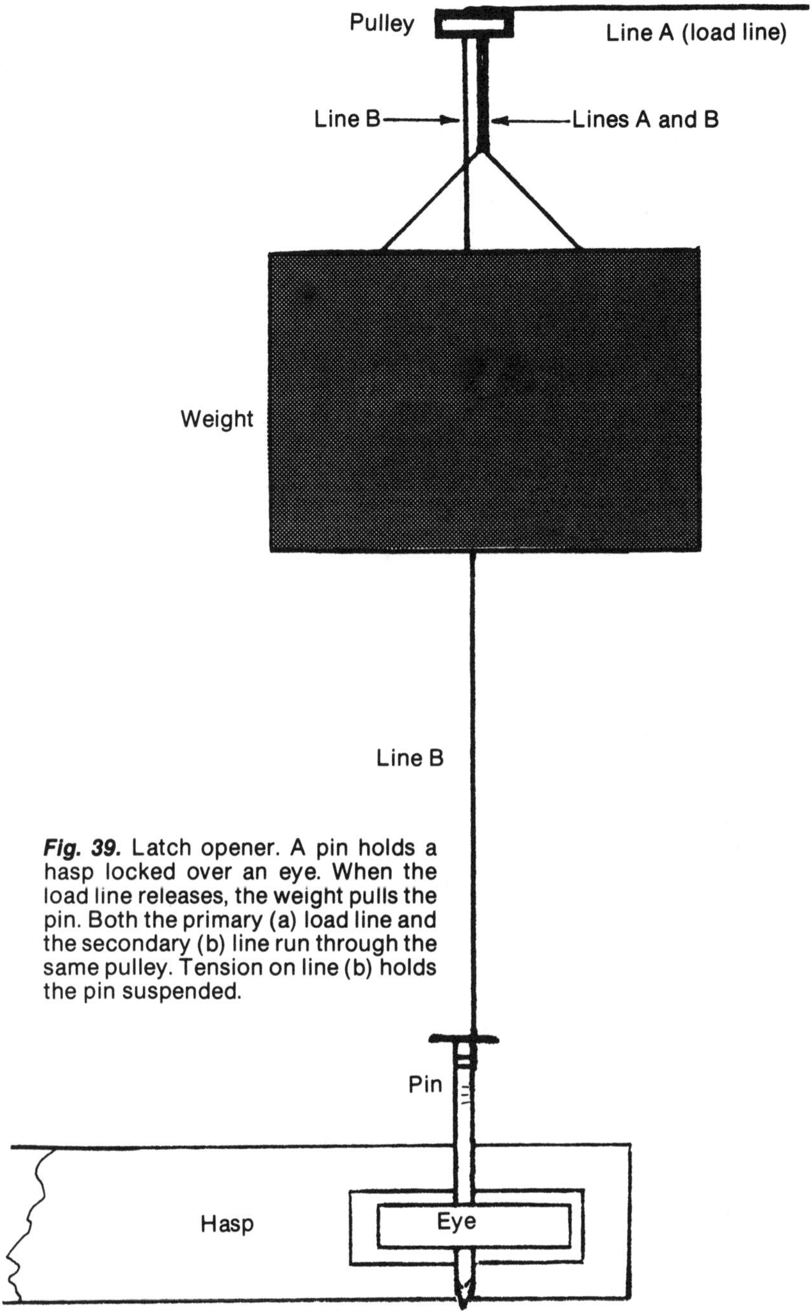

Fig. 39. Latch opener. A pin holds a hasp locked over an eye. When the load line releases, the weight pulls the pin. Both the primary (a) load line and the secondary (b) line run through the same pulley. Tension on line (b) holds the pin suspended.

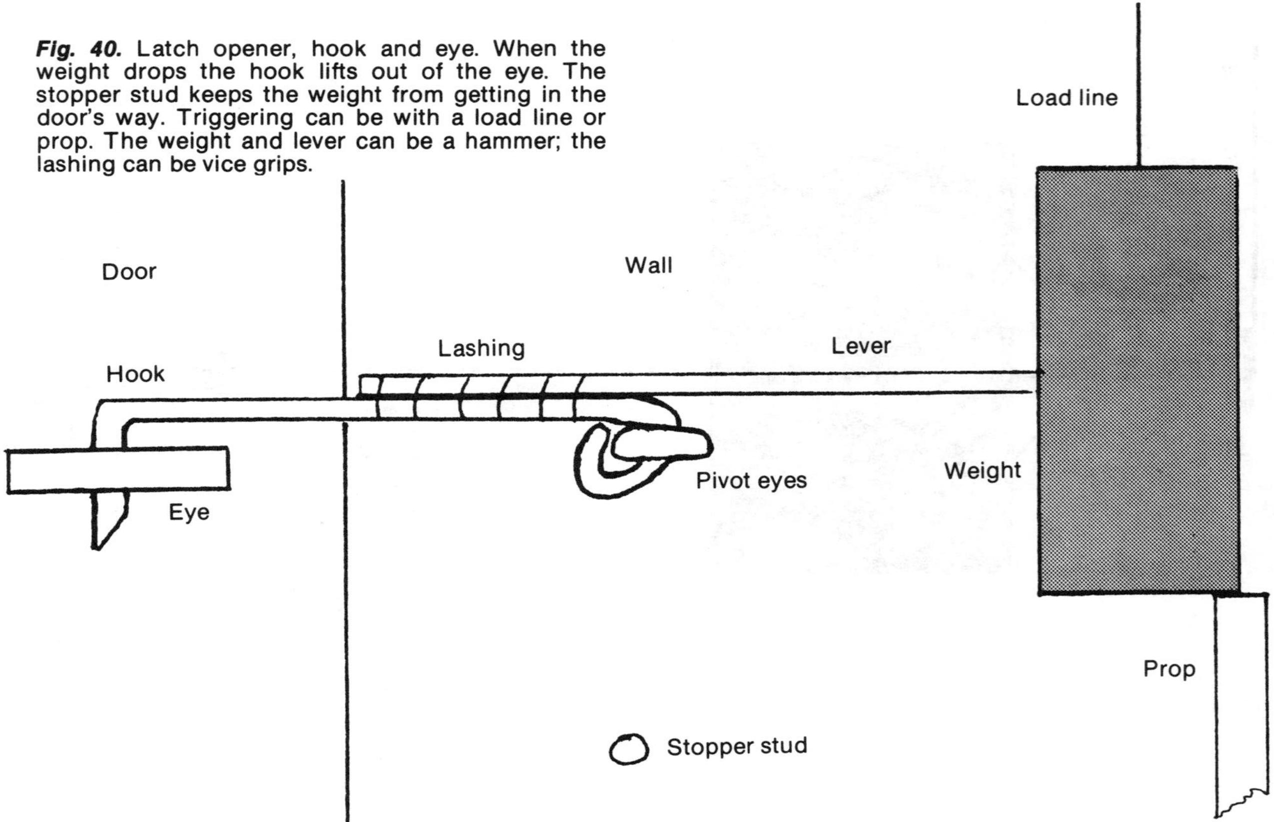

Fig. 40. Latch opener, hook and eye. When the weight drops the hook lifts out of the eye. The stopper stud keeps the weight from getting in the door's way. Triggering can be with a load line or prop. The weight and lever can be a hammer; the lashing can be vice grips.

ing the other side stand up. This can be used as a visual alarm that can leave the target unaware that an alarm has been activated. It can be set in or on a house or other structure.

Smoke Alarm Activator

Most smoke alarms can be activated by a button on the face. The button on a ceiling mounted alarm can be pushed by a falling weight and lever.

Door Bar

A bar can be set to drop down into brackets at either side of a door. This will lock the door more securely than any other locking method, because it braces the door as well as keeping it from opening. It can be coordinated with a door slammer, or used in some other way. The bar can be any strong board, stick, or steel (i.e. angle iron, flat bar). It should fit tight against the door, with just enough slop for the bar to drop easily into the brackets. The brackets must be attached securely to the wall. To be certain that the bar does not bounce out of the brackets as a result of repeated pounding on the door, fingers (i.e. nails) can be set at either end, protruding downward.

Section II

Swinging Deadfalls

CHAPTER 6

Form Structure and Characteristics

A swinging deadfall is connected by an arm to an overhead anchor, which directs the course of its fall. Rather than dropping straight, it swings in an arc; down and in, like a pendulum. This allows it to cover a much larger area than a straight deadfall, which makes timing easier and more certain. Since it attacks from the side instead of the top, aim is also easier, because the target area is larger. Spikes can be pushed in with less effort, because they will penetrate muscle instead of bone. Concealment is easier, because the payload does not sit directly over the target.

This is the most complex of the three forms. It is more difficult to build, and more likely to malfunction if not built correctly. But this minor drawback is more than compensated for by the larger kill zone and greater versatility it offers.

It consists of an arm, a pulley, a load line, and a trigger. The arm is the payload and its delivery system, which can be either hard or soft. The top of the arm is hinged, and sits approximately over the target area. The low end is held poised under the pulley by the load line, which attaches to the trigger. When the trigger is activated, the arm is released, to swing down into the target.

If the trap is set so that the target will be between the payload and a solid surface, such as a wall or tree, effective force will be increased.

CHAPTER 7

TRIGGERING

A swinging deadfall is triggered in the same way as a straight deadfall, with the difference that it will sit in a different place in relation to the payload.

Timing

What this form lacks in simplicity of construction, it gains in simplicity of timing. Providing its course is more or less parallel to the target's direction of travel, tolerance of error is much greater than with a straight deadfall.

This is obviously not the case in regard to targeting moving vehicles. Since the object there is to set a block down that the vehicle will smash into, this form is actually less likely to make a hit than a straight deadfall. Very slow moving vehicles, however, that are attacked in the same way as human targets, can be hit, and the large kill zone applies to them.

The payload reaches maximum speed and power when it passes directly under an anchor, making that point the optimum target area. The amount of time it takes the payload to reach that point is slightly greater than the amount it would take a weight to freefall a distance equal to the length of the arm; so timing is set as though a weight was going to drop straight down from the pivot point. The area in the kill zone in front of and behind that point is tolerance of error.

For more specific information on timing, see Chapter 2.

Hard Arm Prop Trigger

A hard arm can be held up by a prop. The arm can be something that looks like it is supposed to be propped up, like a section of a ceiling, etc. The prop can look stable, but be very unstable.

For more information on props, see Chapter 11.

CHAPTER 8

RIGGING

Rigging consists of a top anchor, one or more pulleys, and a load line. Weight, concealment, good aim, and a good trajectory are the primary factors. There must be enough weight to make the trap effective, and it must be compatible with the rest of the energy system. It must be hidden from the target; and it must hit the target. A good trajectory will enhance power; a bad one will rob power.

The more weight there is, the more difficult rigging becomes. Heavy rigging requires heavy-duty lines and anchors; large objects are more difficult to conceal than small ones; mistakes are more likely to happen, and at the same time are more costly; things like setting triggers become more difficult and so on. But weight can also simplify rigging by compensating for sloppy work. If enough weight can fall in approximately the right way and hit the target somewhere, it will be effective.

The more important concealment is, the more complicated rigging becomes. Lines have to be turned and run behind things, or made to look like something else. If it can be known that the target will be moving through darkness, or in such a hurry as to not be able to look around carefully, rigging is much easier.

A good trajectory means that the payload swings down straight, smooth, and uninhibited. A bad trajectory will cause energy to be lost in the form of the payload twisting and bouncing. The payload may or may not be in the right position when it contacts the target. Incompatibility between lines and pulleys will rob power by causing friction.

Aim

One advantage of this type of trap over a straight deadfall is that it is easier to hit the target, because the payload will cover more area. Aiming is also more complicated, because the payload will swing rather than drop straight.

It is generally better to aim the trap so that the payload swings in the same direction that the target is moving in. In other words, the trajectory should be parallel to the target's path. Trying to hit from the side involves knowing how fast the target will be moving; close tolerance timing, and a certain amount of luck.

The top and pulley anchors act as the front and rear sights on a firearm, respectively. The mechanic should imagine a straight line between them extending to the target. If the payload needs to hit the target in a certain way, it is aligned with these two points. If the payload itself is aimed at the target and the two points are not (i.e. the payload is not parallel to a line between them), it will not travel straight. It is the two points that determine the direction of travel, not the way the payload is pointed. The payload must correspond to them. Fig. 41

The payload will have to change position along its course, either relative to the arm, or to the target, and in some cases to both. This will not be a factor for most hard arms, since they will have a fixed payload. A soft arm payload will hang from the pulley anchor either in the position it will be in when it hits the target, or in the position

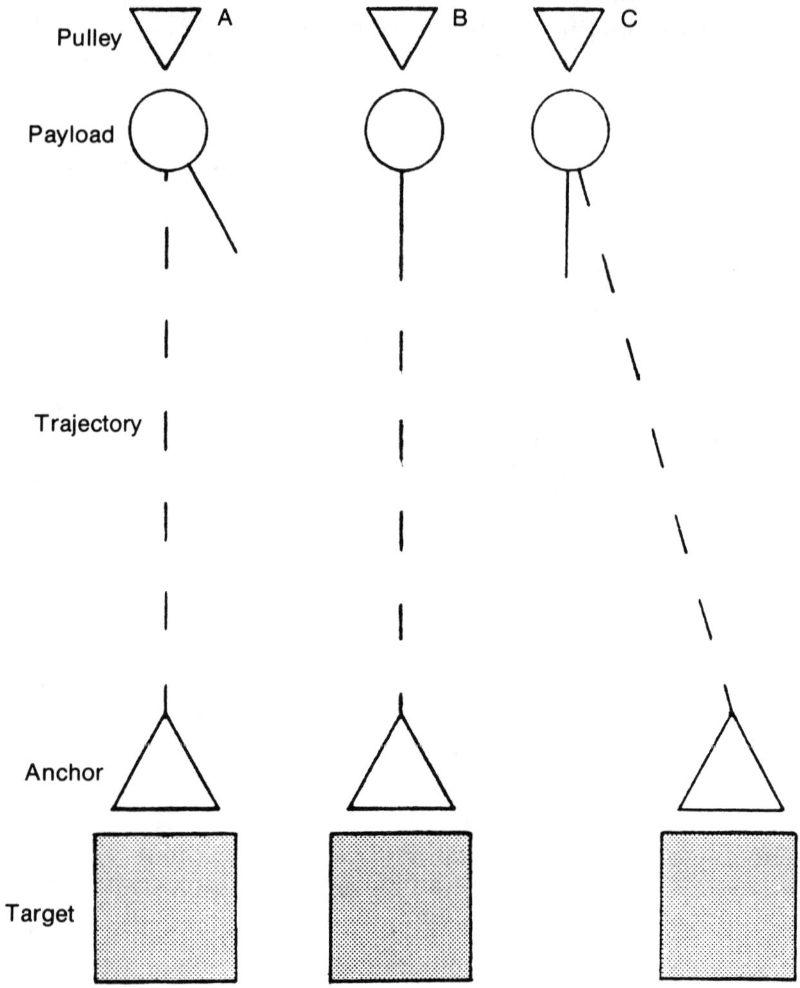

Fig. 41. Aim. Overhead view. The top anchor and pulley determine the course the payload will take. They should be aimed at the target, and the payload aligned with them. In (a) the pulley and anchor are aimed at the target but the payload is not aligned with them, and so is not pointed toward the target. In (c) the payload is pointed toward the target but is not aligned with the pulley and anchor. All three payloads will hit the target, but only (b) will hit correctly.

it wants to set on the arm in, or somewhere in between. Fig. 42

Where the load line is attached to the payload and how much tension the anchor line has on it will determine how the payload hangs. How it hangs will determine how it falls and, along with the arm (anchor line), what position it will be in when it hits the target.

Different payloads fall in different ways, and different arms have more or less control over them. Some should hang straight with the arm, and some straight with the target. As a general rule, a payload that is small compared to the size of the arm should hang straight with the arm; one that is large compared to the arm should hang straight with the target. It is always best to let an arm take at least a few practice swings, to be sure it is hanging correctly.

The Top Anchor

The top anchor is what the arm pivots on. It guides it after the trap is triggered, and so tells it where to strike. It should be placed so as to cause the payload to focus through the target.

The payload attains maximum velocity when its center of gravity is directly beneath the anchor. There will be less power in front of and behind this point. Fig. 43

In some cases two top anchors will be necessary, to keep the payload from twisting, or for some other reason. Usually the payload wanting to twist one way or the other as it hangs at rest below the anchor will be no problem, even if it has to hit the target in a certain way. If it does not matter how the payload hits (i.e. jug of gasoline), twisting is obviously no problem.

If the payload is pointed straight when the trap is set and as it hangs from the pulley anchor, it will fall straight. If it is allowed to swing back and forth a few times, it might eventually begin to twist. But this is no problem,

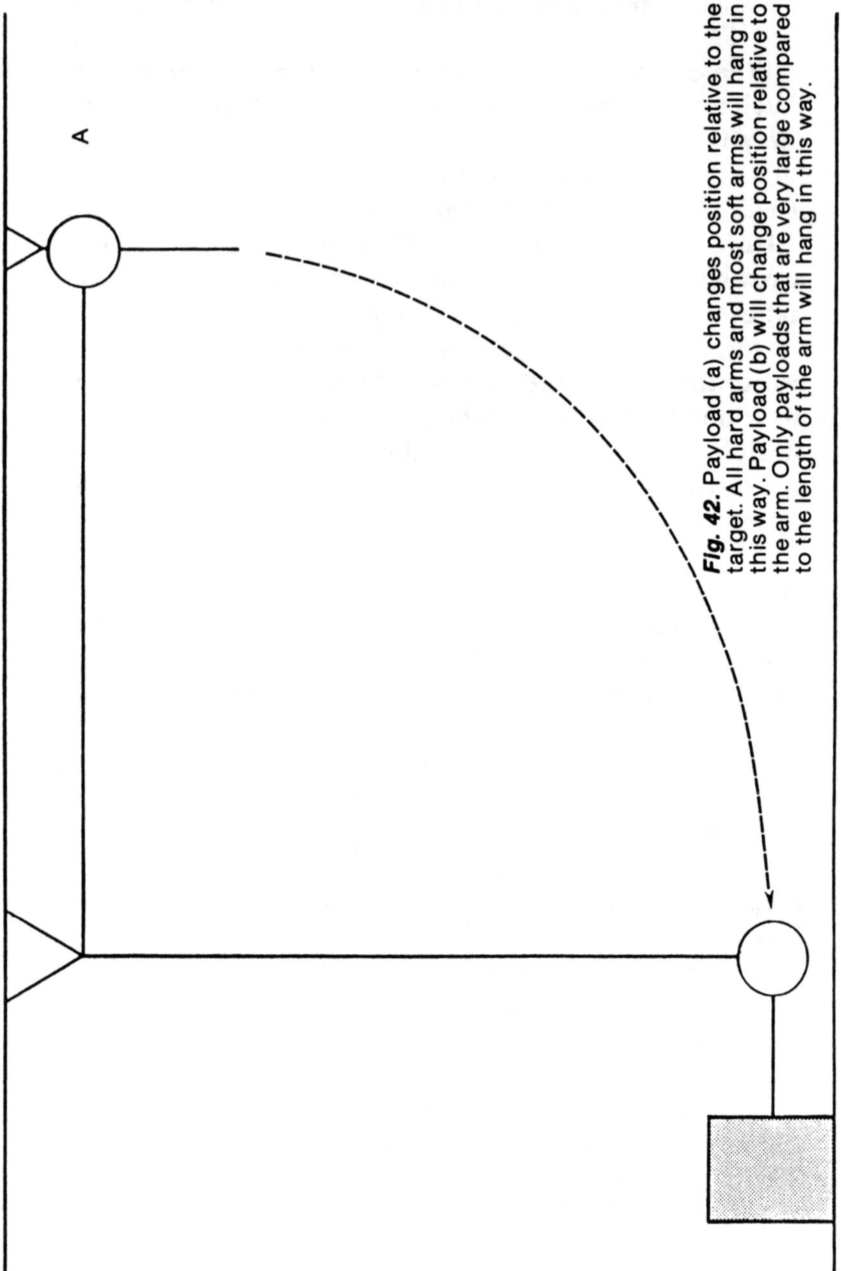

Fig. 42. Payload (a) changes position relative to the target. All hard arms and most soft arms will hang in this way. Payload (b) will change position relative to the arm. Only payloads that are very large compared to the length of the arm will hang in this way.

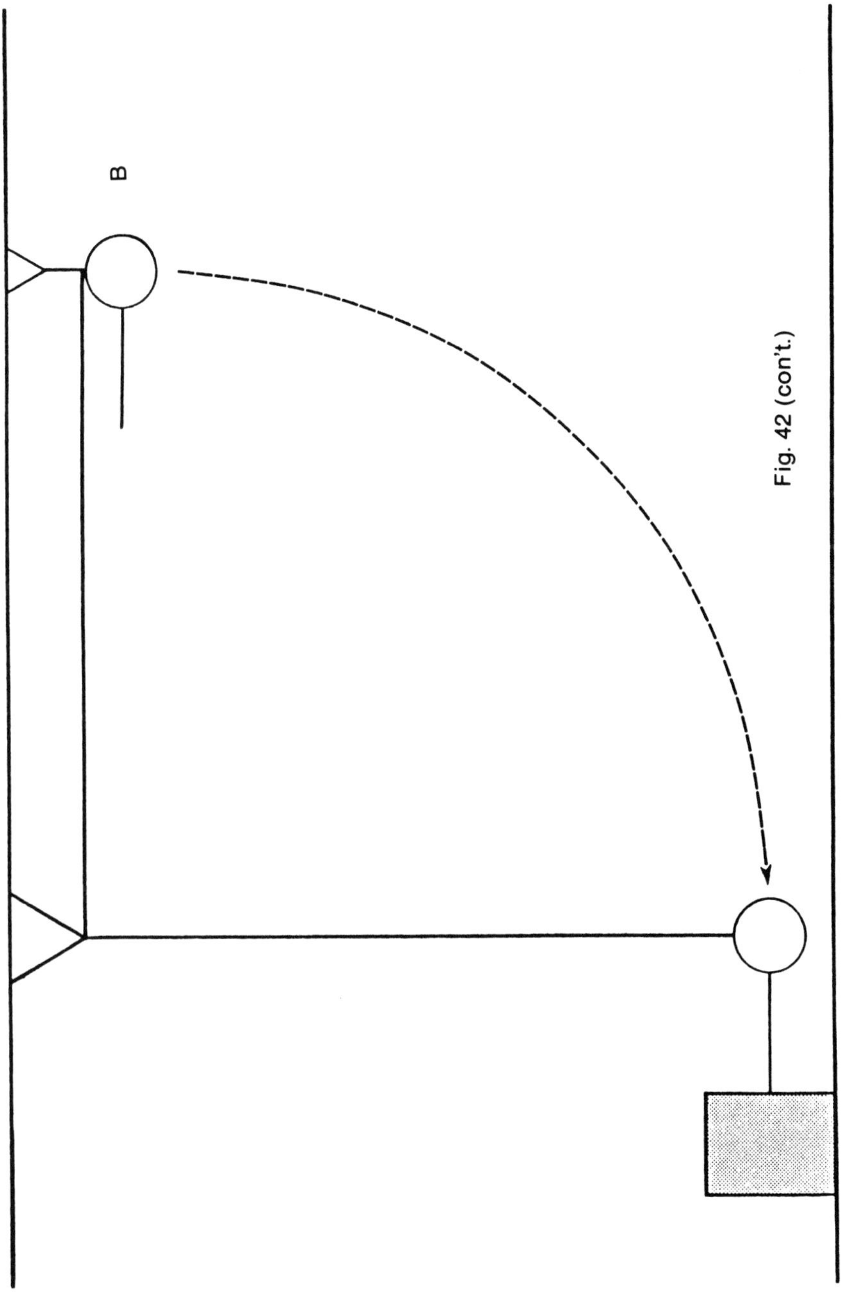

Fig. 42 (con't.)

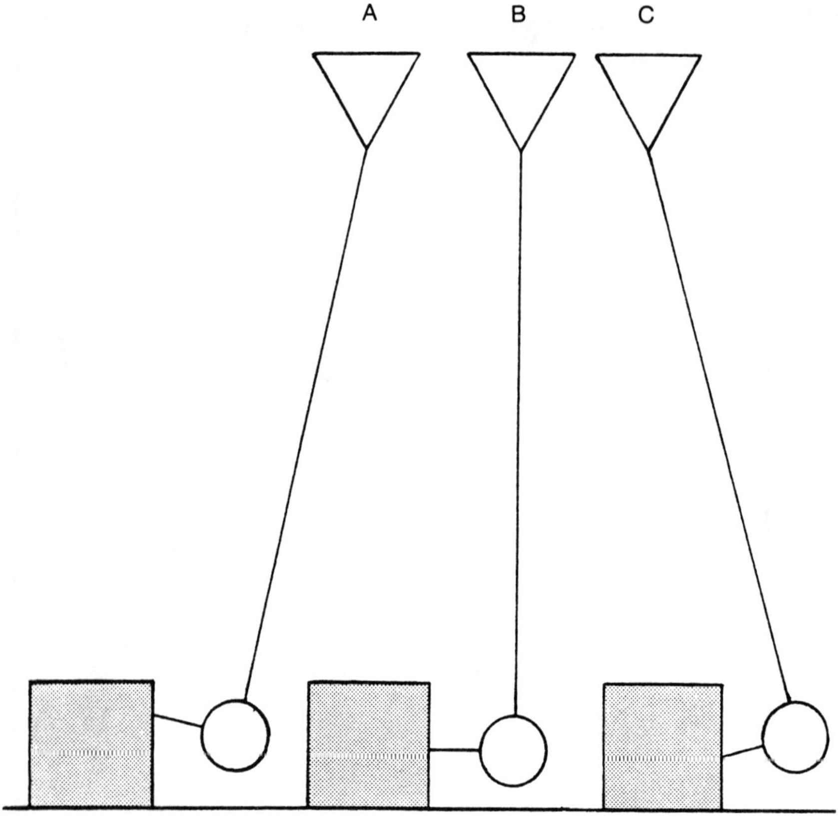

Fig. 43. All three settings are well within the trap's effective range, but (b) will be the most powerful, because it hits the target when it is directly below the anchor. There will be less power in front of (c) and behind (a) that point. Note that in (a) the payload is travelling upward, and in (c) it is travelling downward.

because the first half swing is all that matters. The payload will hit the target in whatever position it is in as it hangs set. This is less true, however, if it travels a long distance (hangs at the end of a long arm), or if the anchor line is very heavy relative to the weight. Otherwise the force telling it to fall straight will be greater than the force telling it to twist, and so it will fall straight. Even very wide or long payloads will usually travel straight with one anchor.

Nevertheless, two will still sometimes be necessary. Two can be an aid to concealment, because there will be nothing directly over the target. Two will support twice as much weight as one. Since the payload will need to be lifted in order to twist or travel off course, the trajectory will be more stable and certain (if the rigging is done correctly). The payload will be less likely to be deflected.

Two anchors complicate rigging, however, to the point that they should not be used if there is not a good reason for it. In most cases one will be sufficient.

When two are used, the payload must be centered exactly between them; and the pulley must be centered as well, so that the trajectory will be exactly perpendicular to a line between the anchors. In other words, the two anchors and the pulley must form an isosceles triangle, and the line of trajectory must divide it into two right angles. If the angles are not as they should be, the trajectory will not be straight and smooth. Energy will be lost as the payload twists and bounces; it might hit the target correctly, but it may not. Twisting is a much greater robber of power when two anchors are used, because the weight has to be lifted.

Anchors can be made from screw eyes, nails, bolts, tree limbs, and anything else that sticks out and has sufficient strength. Holes can be drilled, gouged, or burned through ceiling studs. A line can be toggled into a hole punched in plasterboard. Sometimes pre-existing anchors, in the form of light or plant hangers, will be in the right position.

The Load Line and Pulleys

The pulley acts as a temporary anchor, and as a turning mechanism. It and the load line hold the weight up, charged with potential energy. It turns the load line back toward the trigger.

The higher up the pulley is positioned, and the closer to it the line holds the arm, the more energy there will be, because the weight will have further to fall. Height can also be an aid to concealment. In most cases pulley height will be determined by the position of whatever it is anchored to.

The arm should hang as close to the pulley as possible, with one exception. A soft arm payload should not hang higher than the top anchor. If it does, it will drop and bounce, resulting in a bad trajectory. A hard arm can hang so high as to be almost vertical, and still have a good trajectory. Fig. 44

The distance between the pulley and the top anchor (or the mid-point between two top anchors) should be approximately equal to the length of the arm. This is especially important for soft arms. If it is too close, the anchor line will hang slack, causing the payload to drop and bounce rather than swing in a smooth arc. If it is too distant, the payload will sag down. In addition to making concealment more difficult and robbing power by not allowing the payload to be as high as it would otherwise be, this can result in a bad trajectory. In effect what it does is allow the arm to have more control over the fall than it should, by putting tension on it. Ideally, the pulley should be directly over the weight, and the anchor line should be straight but not tight. Figs. 45 & 46

Actual pulleys as sold in hardware stores obviously make the best pulleys, provided they are the right size and do not hang down too far. A screw eye is almost as good, and is much easier to get, carry, and work with. A fencing staple will work, if it is large enough to accomo-

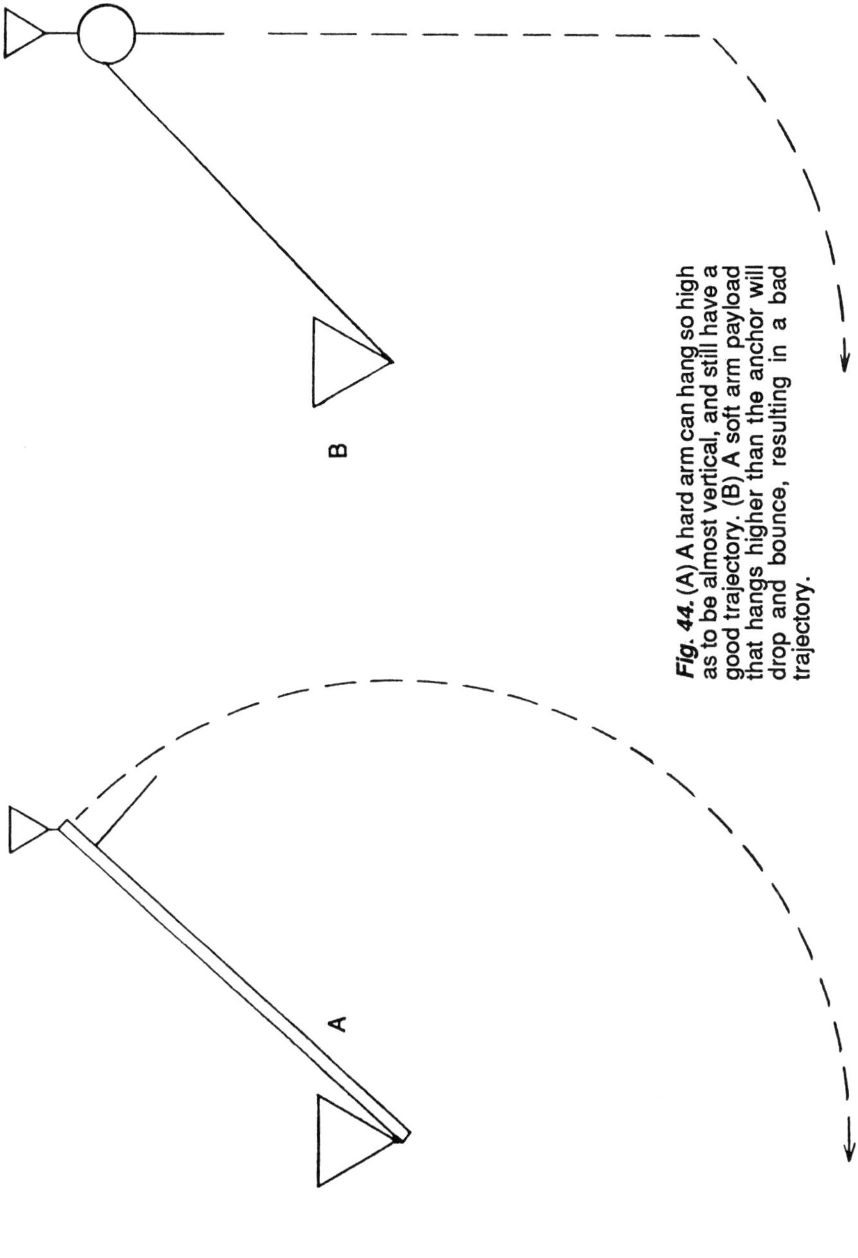

Fig. 44. (A) A hard arm can hang so high as to be almost vertical, and still have a good trajectory. (B) A soft arm payload that hangs higher than the anchor will drop and bounce, resulting in a bad trajectory.

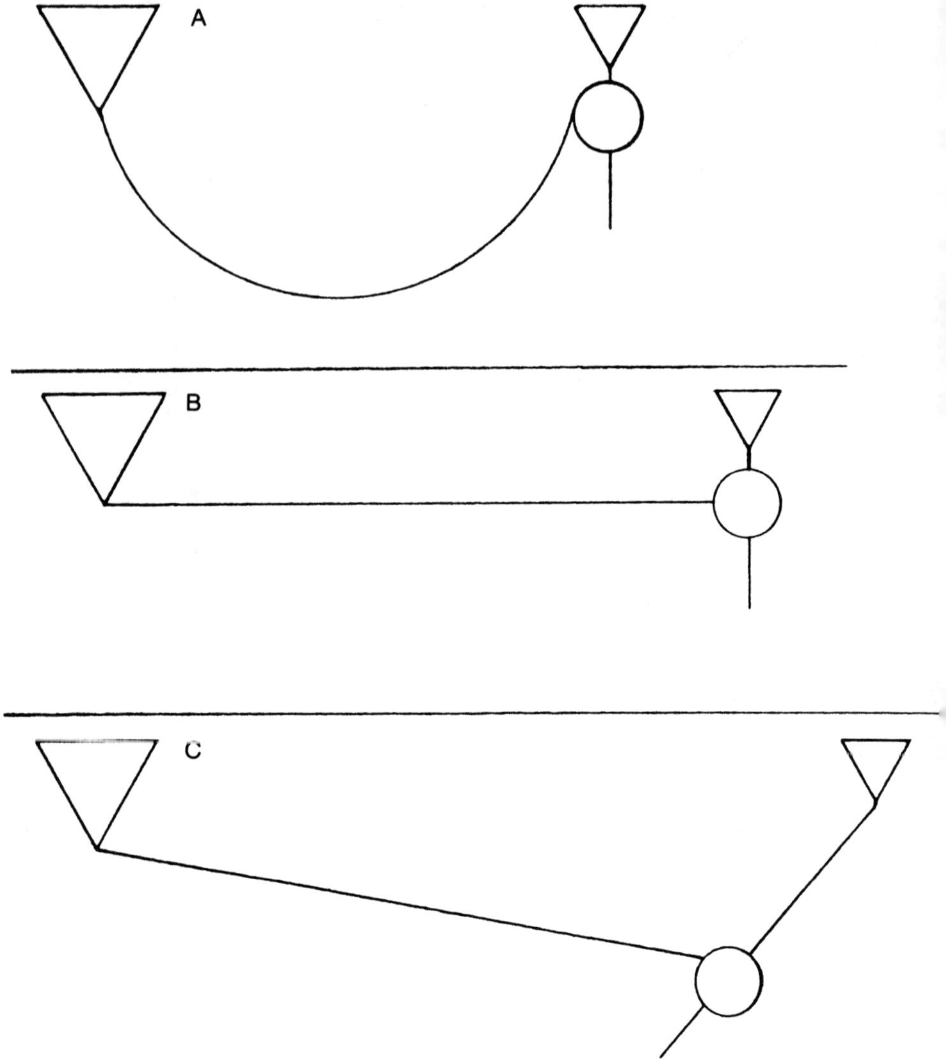

Fig. 45. Soft arm. The pulley should be positioned so that the payload hangs directly below it, and the anchor line is straight, as in (b). Payload (a) will drop and bounce; (c) can not hang as high as it should.

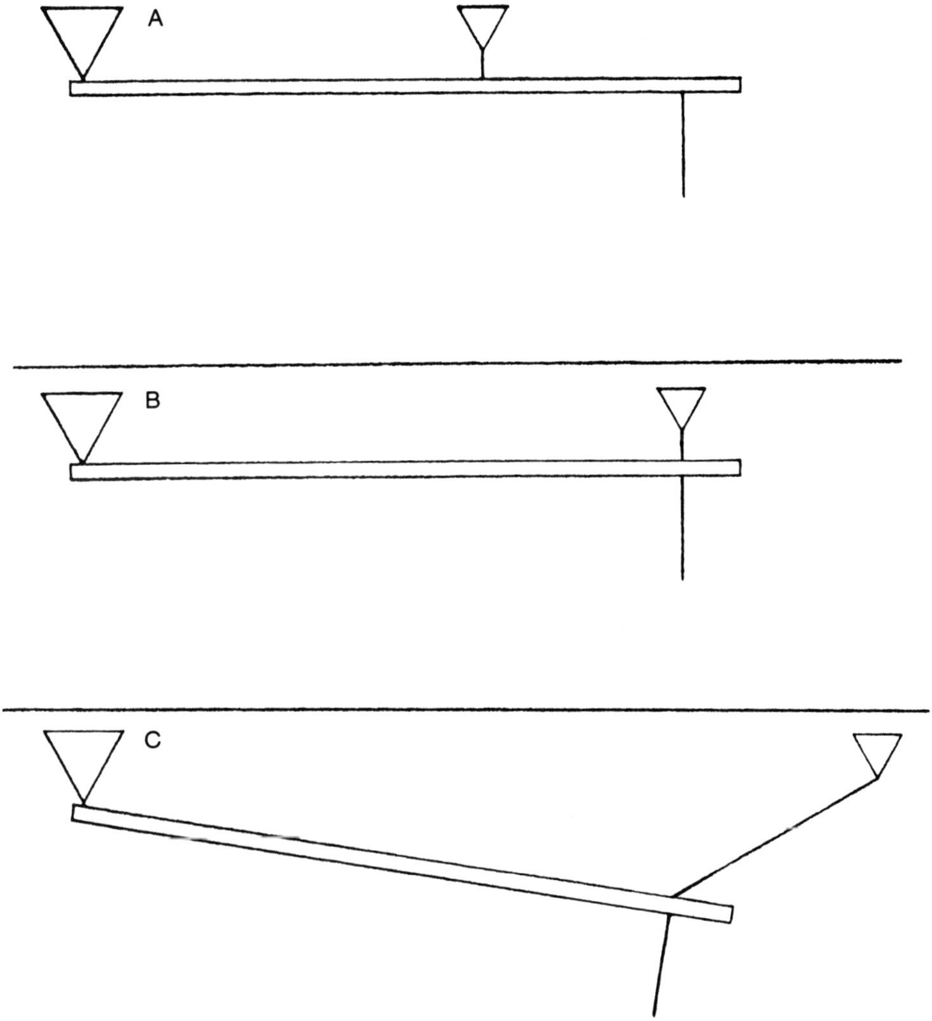

Fig. 46. Hard arm. (a) and (b) are both correct, but the pulley in (a) will have more stress on it than the one in (b). The arm in (c) sags down.

date the line. A nail works well; it may be necessary to bend it to keep the line from slipping off. There is a chance, however, that a loop in the end of a running load line will catch on it (or any other open pulley), and stop or slow the payload short of the target. A hole in a stud makes a good pulley, as long as it is large and smooth enough to allow the line to run freely without snagging. Crotches in tree limbs work well; they might need to be smoothed, either by removing the bark, or by placing something smooth such as tin or aluminum from a can, rawhide, etc., nailed, wired, or tied down between them and the line.

The pulley and load line must be compatible, so that no energy is lost in the form of friction between them. They contribute to power by allowing the motion to be smooth and uninhibited. This is not difficult to accomplish, since the line, once released, has a relatively insignificant amount of weight. The less relative weight and size it has, the better.

Certain substances are not as compatible as others. For example, a rubber line (i.e. electrical cord) will not run over steel as well as it will run over wood. Sisal rope runs over steel better than wood. Nylon runs well over anything. In any case, compatibility can be improved by rubbing graphite, grease, oil, or soap on the line and pulley.

Any knot in the line (for triggering, or connecting two short lines) will make the line less compatible with the pulley. Even a small knot going through a large pulley will take away some amount of power as it hits on the way through, though the amount might be so small as to be insignificant. The angle at which the knot runs through, the relative size of what it runs through, and how many things it hits will determine how much energy it takes away. In other words, the better the rigging is done, the less of a problem knots in the line will be. But it is still an asset for a line to be free of knots and other obstructions altogether.

If the line has to have a large knot in it, and especially if the knot will not fit through a pulley, the length of line between the knot and the pulley must be greater than the distance the payload will travel. Otherwise the payload will slow or stop short of its target. If the knot will not have to run through or hit anything, it will be no problem.

In many cases (i.e. for concealment) it will be necessary to use more than one pulley. Each one will add more drag. But, if done correctly, the total energy drain they cause will be insignificant. Each one is also something that a running load line might find a way to get caught on. So it is best to keep the number of pulleys to a minimum. Rigging should always be kept as simple as possible.

Pulleys should be positioned and angled so that the line does not bind as it runs. The closer the line comes to being straight, the better it will run. It should not be bent around corners unnecessarily.

Since friction is caused by a line dragging over a surface, the more surface area the line contacts, the more energy that will be robbed; so obviously the line should contact as little a total surface area as possible. For this reason, and to minimize the chance of a malfunction, the line should be no longer than necessary.

In order to be kept straight—and concealed—the line might need to be run through holes or notches in ceiling or wall studs. As long as they are large enough, and in a straight line, they will not be a problem. If not, they can add up to a lot of surface area, and many angles. To get them perfectly straight, the line should be held in place, pulled tight, and the areas to be cut or drilled marked.

One way to solve the problem of incompatible rigging is to release the payload at the pulley anchor. Nothing at all is much better than even the best nylon line. This will of course complicate triggering.

The load line must be strong enough to support the

weight of the payload, and should be as smooth and free of knots and other interruptions as possible. Being small and light is a tremendous asset; these qualities can do a lot to compensate for bad rigging.

The line on the load side of the pulley will have more weight on it than the rest of the line, because the pulley will take some of the weight. This means that a small line can be tied to a heavier one just past the point where it passes through the pulley, providing the pulley is large enough to allow the knot to pass through; or a single line can be tied so that two lines run from the load through the pulley, then become a single line. This is done by attaching the line to the arm with a non-running loop (i.e. bowline) large enough that the knot will be on the light side of the pulley.

The best material for the loadline is braided nylon cord. Other types of rope have less strength for their size and weight, and are less smooth. Cotton rope is the weakest, but also the least expensive, and so is often the most common in the form of curtain pull cords, clothes line, etc. For large amounts of weight, steel cable works well.

Since boobytrapping material will usually only be used one or a few times, there is no reason not to use line to very near its capacity. (Remember that the smaller the line is the less it will drag, and the easier it will be to conceal.)

There is one exception to the above rule. The line connecting the payload to the top anchor will have force greater than the weight of the payload on it as it swings, as a result of falling. If it is just barely strong enough to support the payload at rest, a swinging payload will break it. This will render it unable to pull the payload into the target.

Electrical cord, antenna wire, and things of that type will work if nothing better is available. Sheets, clothing, and other material can be cut into strips. If three strips are braided together, they will have strength that is greater than the sum of the parts. Braiding will also make the line smoother and easier to work with.

CHAPTER 9

Payload

The **Arm** is the payload and its delivery system. Sometimes they are one and the same. As the body, the arm is what introduces the trap's energy into the target. It is the physical object that strikes the target. It has two primary forms: hard and soft. Some arms have elements of both.

A hard arm covers a greater area, because it transfers the force of its weight to every point on it. It can increase the force of its weight by acting as a lever. (Consider the way a hammer head seems to get heavier as it is held further away from the hand.) Weight all the way to the free end of the arm makes the best use of leverage in regard to power. It will apply pressure against any part of the arm that contacts the target, and with more force than if the weight was directly behind the point.

Rigging is generally easier on a hard arm than on a soft one. Since it can act as its own straight edge, it is easy to aim, and it is easy for spikes to be aligned with the trajectory. Since the entire body is rigid, it can be controlled very well from a single point (the "ship of the line" concept). It will remain a straight line, no matter what angles of force gravity and inertia put on it.

A soft arm acts as a free weight falling, and not as a lever. All of the weight and hardness are concentrated at the free end. Weight at that point is not transferred to any other part of the arm. Since it covers a relatively small area, more must be known about the shape of the target than with a hard arm.

Even though rigging is less simple on a soft arm, it is not necessarily more difficult. If the mechanic is reasonably skilled, it can be easier and faster to build and set; and if enough can be known about the shape of the target, one point is all that needs to be covered. It can be easier to conceal, material to make it with is easier to carry, and it has the ability to bend towards the target.

An arm can be built that is both hard and soft. For example, consider a board that is hinged at one or two points, and will bend toward the target. Cement blocks tied or wired together end to end can act in the same way. Two or three small payloads, such as clay balls, can hang on the same soft arm.

This type of arm can to some extent conform to the shape of the target. A good fit means that more points are attacked. This can increase the arm's potential effectiveness, especially when the payload is spiked weights. It will make it more likely that at least one good target will be hit; and it can increase shock by causing multiple wounds.

This type of arm, however, can complicate rigging a great deal. Since it is not rigid, the pulley must be directly over the weight, or the arm will sag down. Weight at multiple points requires multiple pulleys, load lines, and so on.

This problem can be dealt with in a few ways. One is to not care if the arm sags a little. If the pulley is high enough, sagging will not be a major problem. The line between two payloads can be what the loadline attaches to, so that both are supported equally. Or the lower payload can sit on a ledge of some sort, balanced so that when the first payload falls, it pulls the second along behind it.

In most cases, the degree to which this type of arm complicates rigging will far outweigh whatever advantage it might have in regard to greater effectiveness. An arm that is strictly hard or soft is recommended for most purposes.

Primitive Forms

Weight and spiked weight are the most primitive forms of payload. This makes them generally the most simple to build, and the most commonly available.

In order for weight alone to be effective, it needs to be very substantial. This means a lot of weight, a long falling distance, or both. Weight without sufficient velocity will be ineffective no matter how great it is, because it will push the target rather than hit it. If set so that the target will be between the payload and something that will not move (i.e. tree, wall, vehicle), heavy, slow-moving weight can be effective. Such weight can also be used to push a target off a ledge, or to attack a vehicle.

Spikes are well-suited to swinging deadfalls, because the payload hits from the side rather than the top. The thorax and abdomen are easier to penetrate than the skull. Spikes aimed at the abdomen have less chance of hitting bone, and so a greater chance of penetrating deep. Low targets also allow more falling distance, and so more power.

Spikes are more or less the same as those described in Chapter 5, except that these should be longer.

Soft Arm, Horizontal Payload

If a payload is long and slender (i.e. log), it will have more power if it travels so as to hit end first. If it travels side first it will cover more area. In either case rigging is simple. Two anchors might sometimes be used with side first payloads, but never with end first.

This type of payload will be made out of logs, steel beams, fence posts, barbells, and things of that nature.

An end first payload will be a kind of spear, and so can be effective even without a great deal of weight, as long as it has sufficient velocity and a sharp end. Since it must fly straight, good rigging is essential.

If weight can be concentrated at the rear end of end first payloads, the payload will have greater reach than it otherwise would. This can be an aid to concealment, because the anchor sits well away from the target area. The payload can travel a good distance through doors and windows.

The best example of this type of payload is a barbell with several weights at one end. With the load line attached so as to allow it to hang level, the leading end extends well forward. A knife or other spike can easily be attached to that end. A clamp should be set behind the handle to keep the knife from sliding.

Side first payloads work very well with spikes. They need only to be placed on the leading edge, because the other surfaces will not make contact. But if the entire surface is covered, rigging will be easier and success more likely, because the payload will hit correctly even if it spins or bounces.

Spikes can be attached to logs or railroad ties by boring holes or making splits in the wood and inserting them. A board with nails can easily be attached to anything. Fig. 47

A steel rod, pipe, etc., can swing down and attack the shins. Attacking low targets allows maximum falling distance, and so maximum power. This can be useful where head room is limited. A broken shin bone will effectively disable an individual.

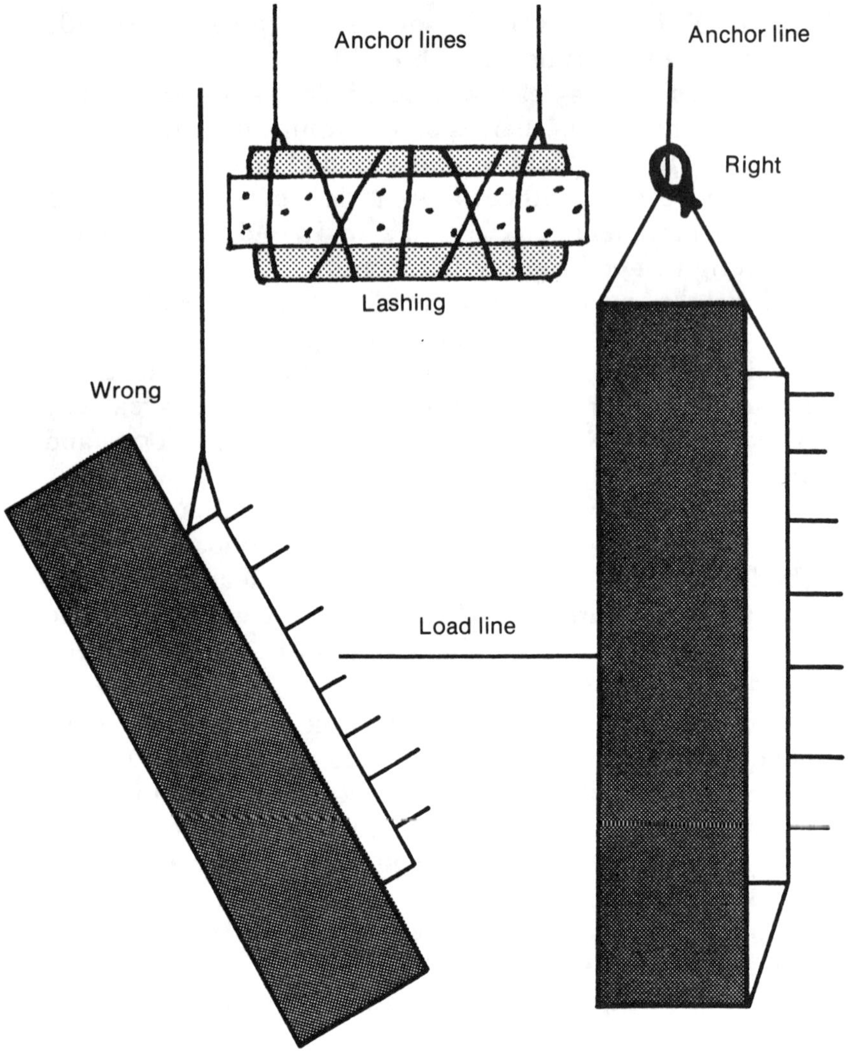

Fig. 47. Weighted nail board. One or two anchor lines can be used. They must be centered over the weight, not the board, in order for the payload to set correctly. The load line attaches to the center of the back. The weight is a bag of dog food, fertilizer, etc. The board should cover as much of the bag as possible.

Soft Arm, Concentrated Payload

A mass concentrated at the end of a soft arm can be very effective. This type of payload includes things such as barbell weights, blocks, stones, and clay balls.

Barbell Weights

Barbell weights can be concentrated simply by running the anchor line through the holes and tying it to itself. The load line runs through the same holes. Or a short bar can be loaded full of weights with a clamp at either end. If the anchor line leads off center, and the load line is attached to one end, the other end will act like a dull spike. Weights used this way can effectively attack a target that is against a hard surface.

Knives with finger guards are very easy to attach to weights, simply by slipping the handle through the hole so that the weight will push against the finger guard. The knife should be secured in the hole so that it sits exactly perpendicular to the weight, and continues to do so after it contacts the target. This requires a good, snug fit. It can be done by wrapping cloth (i.e. hankerchief) around the handle before inserting it. Glue or other lashing can also be used.

The anchor line is attached through the hole in the weight, and the anchor line is tied to the end of the handle (Fig. 48). A hole in the handle simplifies rigging. In order for the knife and weight to sit correctly (perfectly horizontal), the anchor line must lead off the center of the weight. If more weights are used, the line still must be centered. Multiple weights should be lashed together.

The knife should be set with the edge up, so that it will cut as the knife moves along its curved course.

A knife, screw-driver, scissors, or other spike can also be attached to the flat side of a weight.

This is done with tape, wire, or other lashing. It should be aligned with the weight as much as possible, so that if

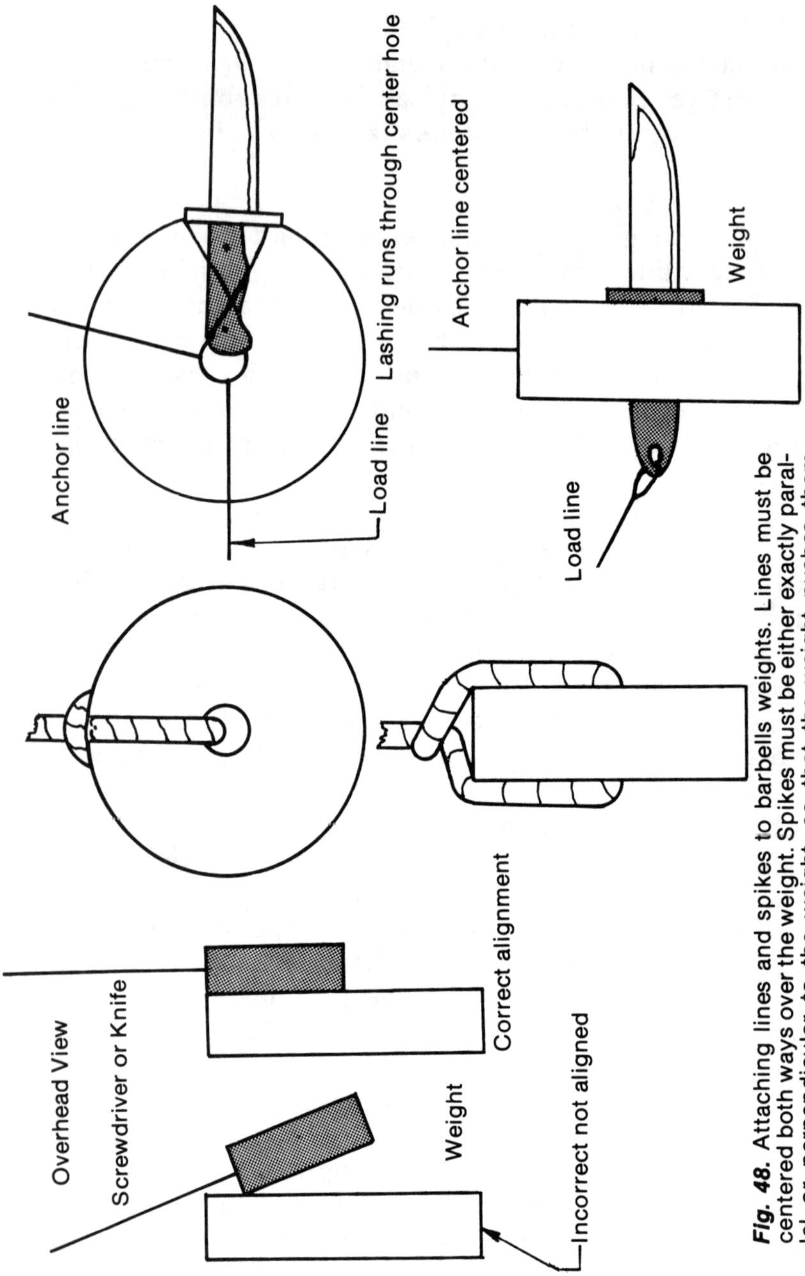

Fig. 48. **Attaching lines and spikes to barbells weights. Lines must be centered both ways over the weight. Spikes must be either exactly parallel or perpendicular to the weight, so that the weight pushes them straight into the target.**

the weight travels straight, it also will, and the weight will be fully behind it.

Knives and other spikes can be attached to other symmetrical weights, such as building blocks, bricks, and books. The same rules that apply to barbell weights apply in every case: the weight and spike must be aligned; the spike must be securely fixed to the weight, with part of the weight pushing against it if possible; the load line must be centered, to allow the spike to sit level.

Clay Balls

The classic example of a concentrated weight payload is a spiked clay ball. If it is made with spikes sticking out in every direction rigging is simple, because it does not matter how it hits.

The ball will be either spherical or cylindrical, and should be as perfectly round and balanced as possible.

The traditional spike used with clay balls is sharpened bamboo sticks. Other spikes include sharpened hardwood sticks, long nails, sharpened welding rods, (preferably stainless) and so on. They must go far enough into the clay to be firmly anchored. They should protrude straight out of the ball, as though they radiate from center. Fig. 49

A much easier way to make a clay ball is with a mold. This can be made out of cardboard, or something similar to a large coffee can. Spikes are set in place through the sides of the mold, and clay is packed in around them.

A mold can also be filled with cement. Cement is easier to work with than clay, sets up harder, and weighs about one and a half times as much.

Hard Arms

A hard arm payload can cover an area equal to the length of the arm. Weight at the low end will apply force all along the arm. While a hard arm will usually be more difficult to hide than a soft arm, it can sometimes be easy

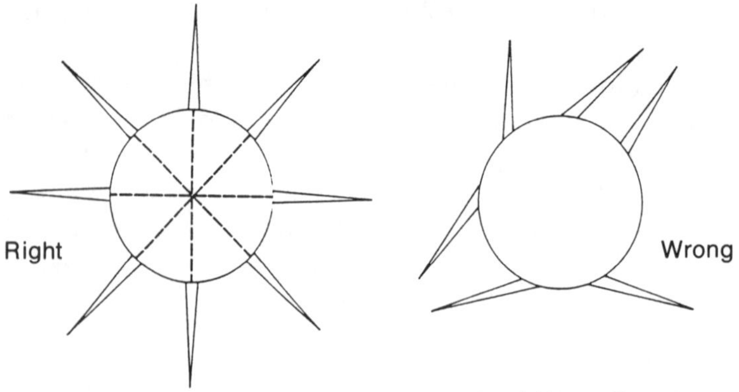

Fig. 49. Clay bell, spike replacement. Spikes should be well anchored in the ball, and should stick straight out, as though they radiate from center.

A payload can be hidden by making it look like something else. In this case a spiked ball is disguised as a bird nest by long grass woven through the spikes. Note that the load line is still conspicuous.

to conceal by being made to look like something else. For example, a pipe hanging at a basement ceiling can look like plumbing. Clothing hanging on a hard arm can make it look like a hanger, and also hide weight and spikes at the low end. A log set in a tree can look like a branch. In some cases a hard arm can be easier to hide than a soft arm, because it can hang so high as to be almost vertical, and so be out of the target's field of vision.

Anything that is long, straight, and rigid can be used as the arm. This includes logs, boards, pipes, rods, I-beams, angle iron, flat bar, and other similar objects.

Weight should be concentrated at the bottom as much as possible, even if a different part of the arm is intended to hit the target. Spikes placed anywhere along the arm will be pushed into the target by the weight. They should be as low as is practical, because the lower they are the faster they will be going, and the straighter their trajectory will be. But hitting a target is the primary consideration.

The anchor end has some sort of hinge mechanism. It should be of good quality and design, because a bad hinge can be as great a robber of power as bad rigging. One of the easiest hinges to make is an eye that pivots around a stud. The eye can be a screw eye, bent nail, or hole in the arm. The stud can be a nail, bolt, or tree limb. The eye can be in the arm or in the anchor, and the stud on the other member.

If no stud and eye material is available, the top end of the arm can be lashed to any anchor, with the lashing acting as a hinge. There must be enough distance between the arm and anchor to allow the arm to swing free and uninhibited, but otherwise the space between the two should be as short as possible. As the space increases, it adds softness to the arm, which will significantly reduce the force that the low end applies against the rest of the arm.

The lashing is done by tying a timber hitch or clove hitch to the arm, and then tying it to the anchor. It is the length of line between the arm and anchor that must be kept short, not the actual distance between the arm and anchor. The closer to the top of the arm this line is tied, the shorter this length can be. Lashing can be done in other ways as well.

If the arm does not want to sit correctly when it hangs at rest, it should be turned in the lashing until it sets correctly.

Ax

An ax is a ready-made lever arm with a combined weight and spike (edge) at the power end, and a bulge to keep things from slipping off at the grip end. It does not have either enough weight or length to be very powerful, but can inflict damage if sufficiently sharp and allowed to fall as far as possible, which means being almost vertical when set. As with any hard arm, power can be increased by adding weight or extending length. Length is extended by lashing a long straight stick to the handle. Adding weight is difficult, because there is not a good surface to set things on. Any added weight would have to be perfectly balanced and in line with the edge. But if this much work is going to be done, it might be better to use the ax to make a good arm out of something else.

It is very important that this payload be traveling perfectly straight when it hits, or it will turn sideways instead of cutting.

Nail Board

A board with nails or some other spikes attached can be both easy to make and effective. Boards are easy to work with, because they are made for the purpose of building things.

The easiest way to hinge a board is to drive a nail

through the anchor end into a wall or beam and let the board pivot on it. In this way the arm sets flat along a wall. The nail should be loose enough to allow the board to swing uninhibited. It can be made loose by pivoting the board on it several times. The board should not be so tight against a wall that friction robs power.

An easy way to attach nails to a board is to drive them through a thin board and then nail that board to the hinged board. Or nails can be driven part way in and the heads cut off with a saw, file or bolt cutter. The nail is cut at an angle, and then filed or ground to a good point.

Weight can be lashed or nailed to a board. A wide board nailed to the arm makes it easy to attach weight by giving the weight something to sit on. Weight can be bricks, bags filled with steel or other metal (i.e. nuts and bolts, coins, fishing weights), steel rods or pipes.

A board can be hung so as to swing wide side first by making a bent nail or other hinge in the top. In this way nails for spikes can be simply driven through the board. Additional weight can easily be lashed or nailed on against the back side. Fig. 50

Barbell

A barbell with all of its weight concentrated at the bottom can hit with a great deal of force. If the target will be against a wall when hit, no spike should be necessary.

This arm will be hinged with lashing. A clamp is used to keep the lashing from slipping off.

The target can be attacked at any point along the shaft by clamping one large weight at the point. Or the mass at the bottom can be the striking area. Fig. 51

A knife or other spike can be clamped, glued, etc., to a weight anywhere along the shaft. It is set so that it pushes against the bar, and clamped between a weight and clamp. A pilot's survival knife works well for this because of its flat-sided pommel. Other knives can be clamped

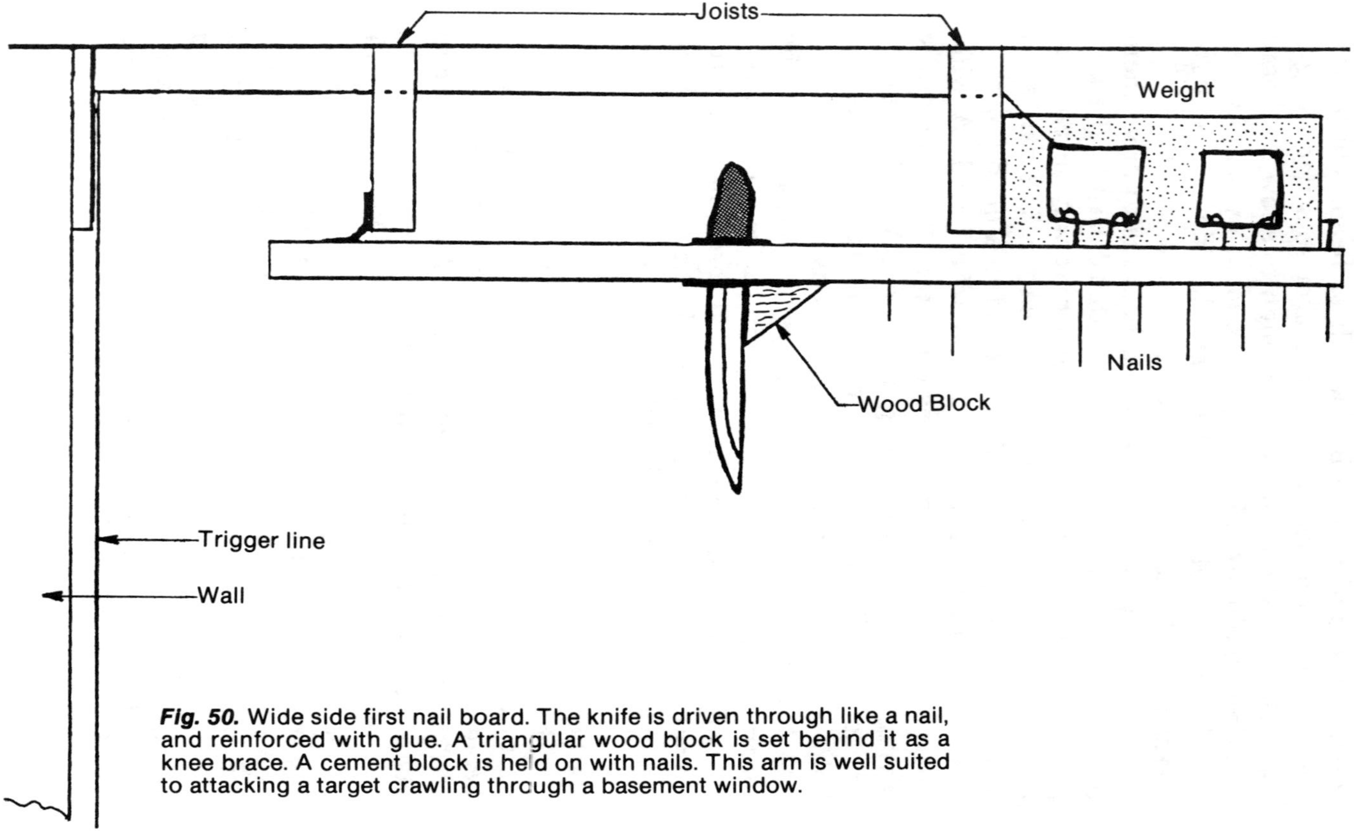

Fig. 50. Wide side first nail board. The knife is driven through like a nail, and reinforced with glue. A triangular wood block is set behind it as a knee brace. A cement block is held on with nails. This arm is well suited to attacking a target crawling through a basement window.

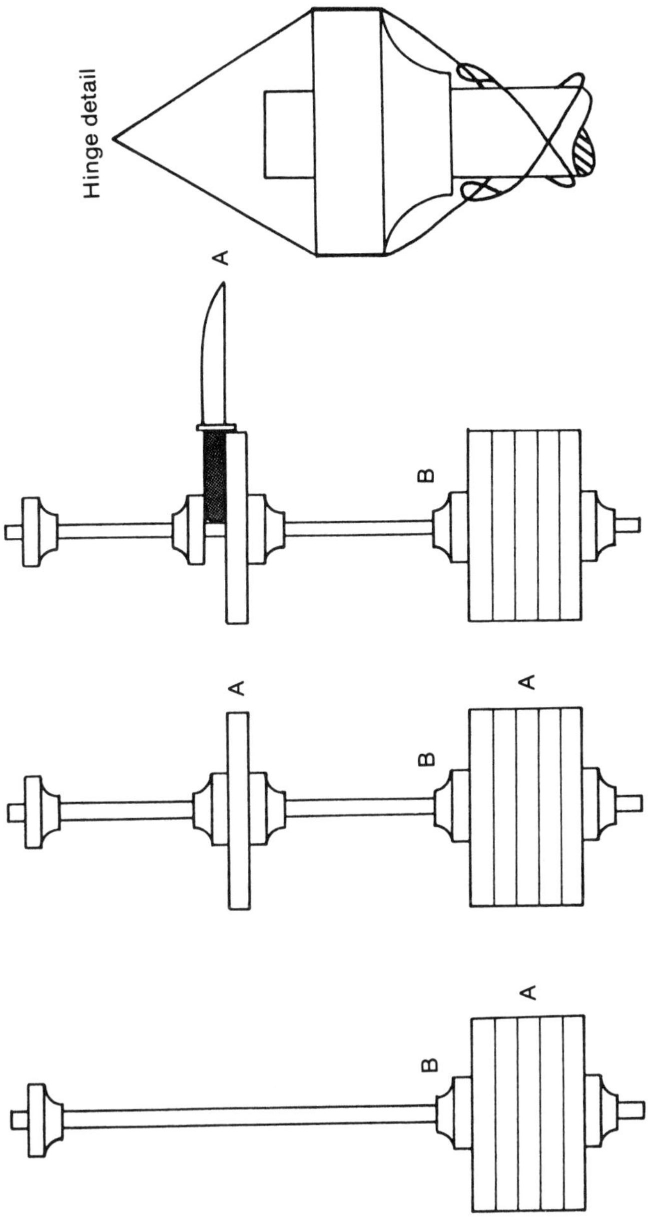

Fig. 51. Barbell hard arm. (a) is the point/s on each arm designed to hit the target. The knife is a pilot survival knife, set to cut upward. The clamps and weight hold it securely, with the weight acting as a brace. It must be set so as to drive perfectly straight into the target, or the arm will rotate instead of pushing it in. (b) clamp is optional in each case.

and/or lashed on in other ways. It should be set to cut upward. The weight it sits on will act as a knee brace.

Logs and Sticks

There are several ways that logs and sticks can be hinged to branches and other things (Fig. 52). Weight can be added by lashing a log or rock to the bottom, or by molding clay around it. Sharpened sticks or other spikes can be lashed along side of the arm, set in holes bored in it, or stuck in clay.

Construction Steel

Construction steel such as I-beams make very good hard arms because they are both heavy and easy to work with. When construction steel is available, tools to shape it with, such as torches, welders, grinders, and drills, are usually close by. Without such tools, it is not easy to work with.

An arm can be made from a piece of angle iron that can be assembled and disassembled quickly and easily. One hole is cut or drilled in the top for a hinge, and more are cut along the arm for spikes. Additional holes can be cut so that additional weight can be bolted on. The angle should be as heavy as possible, which means being as thick as possible.

Spikes are made by grinding points on bolts, which are set in holes and held in place with nuts and washers. They should be long and thin. The bolts must be threaded nearly all the way to the head, such as machine screws, all thread, and others.

The arm is easily disassembled by removing the nuts and bolts, and just as easily reassembled. Care must be taken when grinding points on the bolts to preserve the integrity of the threads. Some work with a file might be necessary before a nut can be started.

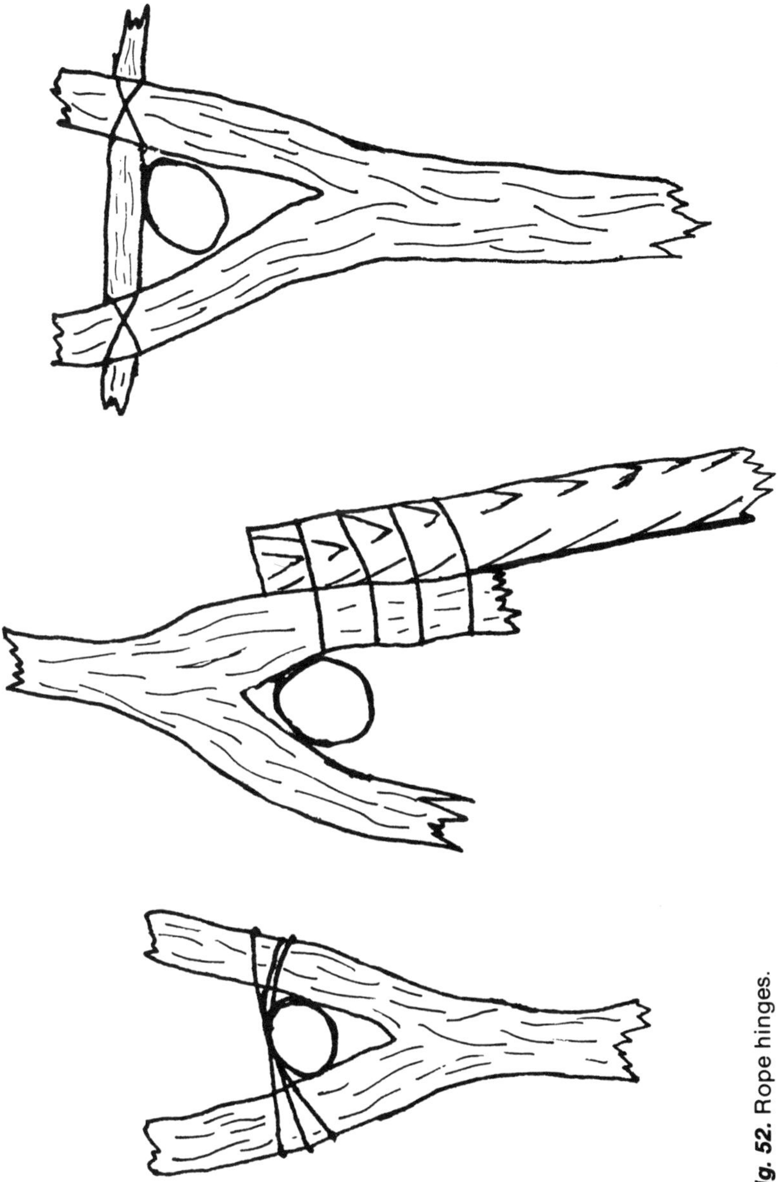

Fig. 52. Rope hinges.

Platform

A hard arm sitting horizontally is a platform on the top side. When it swings down it is a trap door. Any hard arm can have things sitting on it to drop straight down as the arm swings. This can increase the size of the kill zone.

An arm can also be nothing but a platform, with several payloads sitting on it to drop straight down. It can be hinged the same as any hard arm. Or a ledge on the arm can support the hinge end, which will allow the arm to swing part way and then drop. Such a ledge can be very strong, and will allow the platform to sit perfectly flush with the surface that it is a part of. A hinge that will keep the platform from dropping free, and so cause it to swing, can easily be added with lashing.

The payload that sits on the platform can be anything, as described in Chapter 5. A bucket of fuel with one or more igniters can be very effective on a platform.

A platform can be made to look like part of a ceiling. Concealment is very easy, because the payload and rigging are above the ceiling, and so out of the target's field of vision. It can be painted over if necessary.

A rock fall is a platform. This can be a very effective payload in mountain areas, using available resources. The platform is hung or propped the same as any other hard arm. No hinge as such is necessary, because the platform itself might as well fall along with the rocks. It should be set well above a trail where the grade is near vertical. If the trail is close to a cliff, the grade can be less steep, because the rocks can push the target off the cliff. Given the right conditions, a rock fall can start a land slide.

Match Striker

Strike anywhere matches are attached to the head of a hammer, with the tips at the furthest foreward point. It is mandatory that the tips be struck rather than smashed. This is accomplished by causing the hammer to drive in at approximately a forty-five degree angle. The driving force hitting at an angle together with the hammer wanting to twist after it hits will strike the matches. Fig. 53

The striking surface should be clean, dry, and as much like a factory-made striking surface as possible. One can be removed from a match box and glued or taped to a surface. Fine sandpaper or emery cloth will work very well. Smooth cement works well.

A good arm can be made from a carpenter's claw hammer. Anything with sufficient weight and size that matches can be attached to will work. Too much weight might smash or break the matches instead of striking them.

The matches are attached (glued or taped) securely to the hammer. Since matches are not exceptionally dependable, at least three or four should be used, all set so as to hit the striking surface simultaneously. This is done by setting them in a straight line vertically. When the hammer hangs at rest, each one should touch the striking surface. They should be on the side of the hammer that is closest to the surface.

Good rigging is very important, because the hammer must drive straight and hit at the correct angle. Bad rigging will result in less certainty of ignition.

Fire is transferred from the igniter to the fuel or explosive in a variety of ways. A fuze of some sort can sit close to the contact point. The hammer will swing and bounce right after it hits, but will eventually come to rest at that point. Match heads glued or taped directly behind the contact match heads will ignite, and increase the size and duration of the flame. Matches attached to the striker surface close to the contact point can act as a fuze to lead fire to the next medium.

The fuze can be a vessel of liquid fuel with a wick extending to very near the contact point. The wick should not touch the surface, or it might contaminate it with moisture.

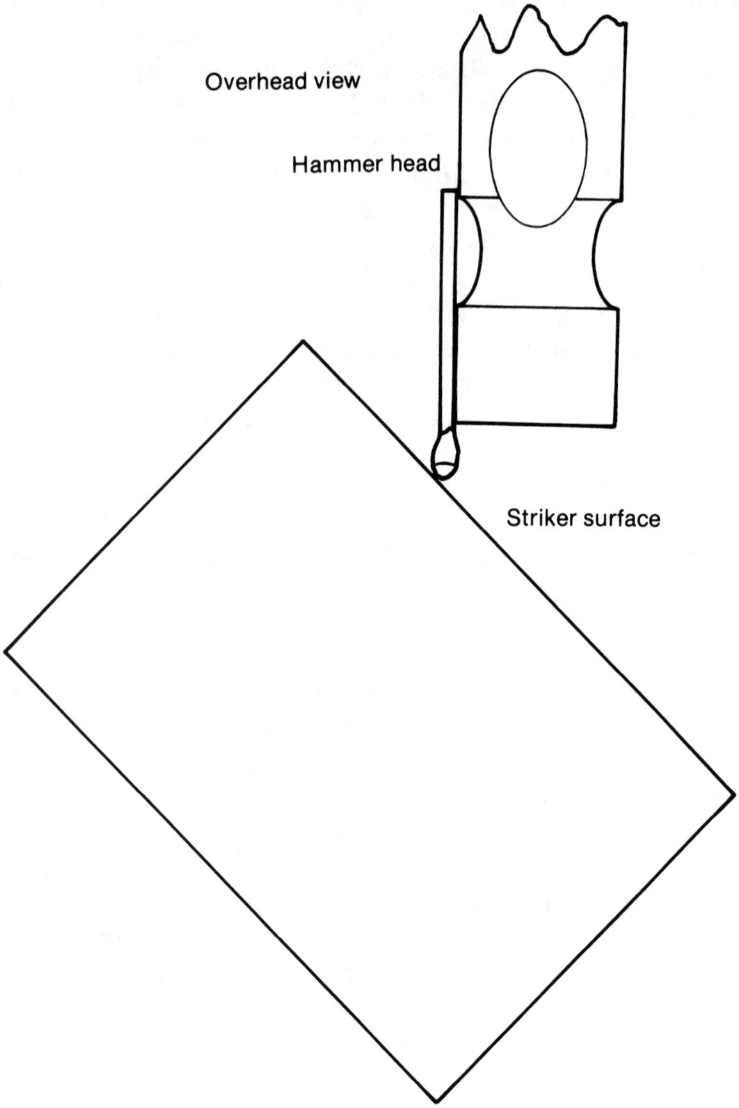

Fig. 53. Match striker. Match heads form a straight line vertically, so that they hit the striker surface simultaneously. They drive in at an angle, so they are struck instead of smashed. The matches are on the side of the hammer closest to the striker surface.

A fuze can be attached to the hammer head itself in several ways, with the end set close to the match heads.

Firearm Hammer

A hammer can be set to swing down and drive a firing pin into a primer. This can initiate fire or explosion, be an alarm, or fire a projectile. The elements needed are the firing pin, a housing for the pin, and a hammer.

A good firing pin can be made from a nail. Anything with a correctly shaped end that can be set in a housing can work. The end should be as similar to a factory-made firing pin as possible. Looking at the dent in a fired primer will give a good idea of what the best size and shape is. Nail ends should be cut off and/or rounded by filing, rubbing on cement, etc.

The housing is anything that can hold the pin poised centered over the primer, and allow it to push into the primer when hit. This can be something such as wood or soap with a hole drilled in it, or something moldable like clay.

The hammer needs to have sufficient force to fire the primer. A carpenter's claw hammer works well and is easy to work with. One can be used to make the trap, then used as part of it. Common red bricks work well; bricks with holes simplify rigging.

This firing mechanism can be employed in a variety of ways. One example will be given in detail.

Twelve Gauge Pipe Gun

To have any power at all, a cartridge must be set in a chamber, and should have at least some barrel. Shotgun shells especially need some barrel length, because they use slow burning powder. A steel pipe as sold in hardware stores makes a very good chamber and barrel. A twelve gauge shell requires a three-quarter inch ID (inside dimension) pipe. One end should be threaded to receive a pipe cap.

Pipe length will determine power, range, and shot spread. Short pipes yield low power, short range, and much spread. As length increases, power and range increase, and spread decreases. A four inch pipe will have an effective range of only a few feet. As length increases, range increases dramatically. A six to fourteen inch pipe should be suitable for most situations.

Steel pipes usually have a seam on the inside and a burr around the mouth, both of which need to be filed off before a shell can be inserted. This can best be done with a small round file.

A hole just slightly larger than the firing pin is drilled in the center of a three-quarter inch pipe cap. The hole must be well centered, so as to be perfectly in line with the primer. The firing pin (nail) is inserted through the hole and held centered as moist clay is packed loosely around it until the cap is filled almost to the top. The remaining threads should be kept clean. The end of the nail should protrude from the top of the clay.

An empty shell with the primer removed is set in the pipe. The cap is then screwed on, with the nail extending through the primer hole. As the cap is tightened the clay is packed, and the firing pin is set securely in the right position. The pin is carefully twisted back and forth until it can slide in the housing with little effort. There should be no side to side movement. Once packed tightly, the clay will hold its shape. Since it is just slightly moist, there should be no significant shrinkage.

The cap is then removed, and the empty shell replaced with a live one. The firing pin can be removed for safety, and reinstalled when the trap is armed.

Fig. 54 shows detail and correct alignment.

The easiest way for the gun to be mounted is for the entire mechanism to be built on a wall or large, flat board.

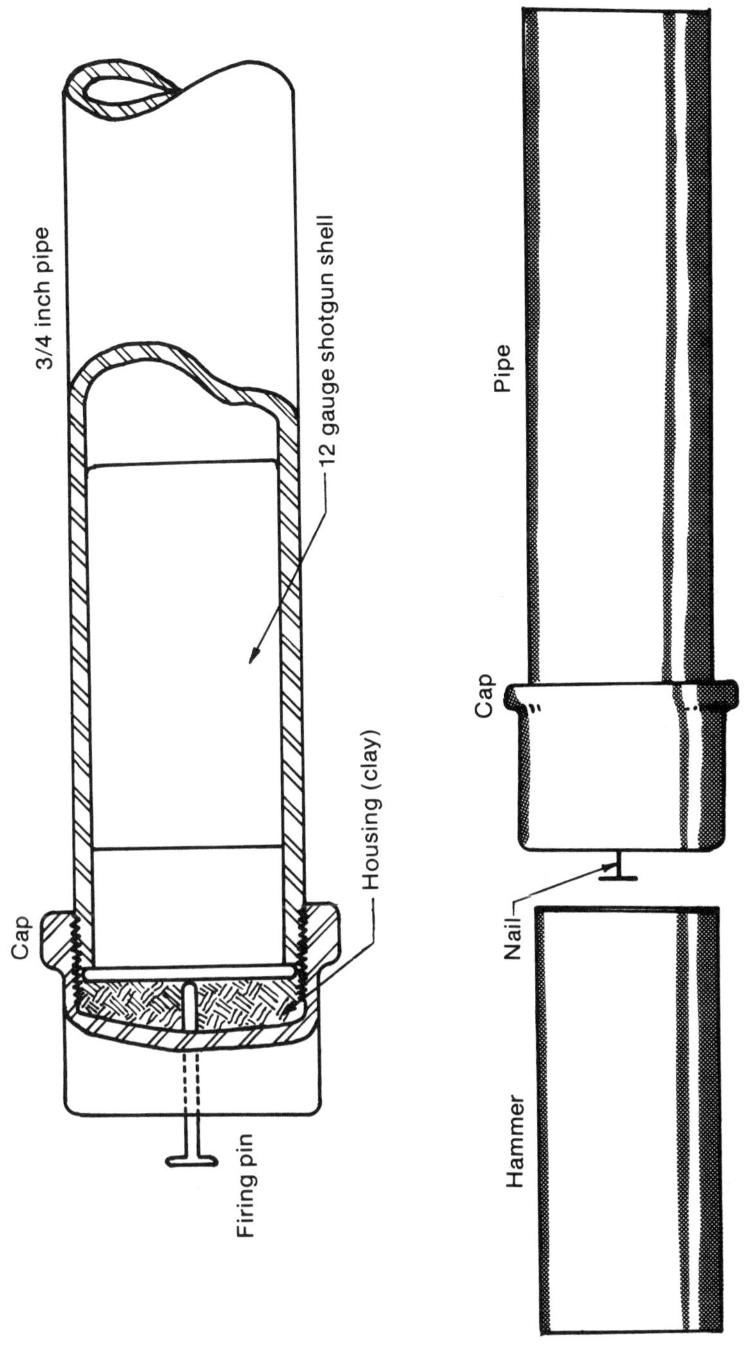

Fig. 54. Pipe gun, assembly detail and correct alignment.

Nails are driven in along both sides of the pipe, with two directly in front of the cap, and then bent over the pipe. Another nail is driven in against the back of the cap to secure against backward movement. This nail should not be set so close to the firing pin as to inhibit its movement. Fig. 55

The hammer is hung so that the handle is vertical when the head touches the nail. It should not be in such close contact with the wall that its swing is inhibited. A hinge can also be made with tape.

Aim can be adjusted slightly by placing a spacer between the muzzle end and the wall. The angle away from the wall can not be too great, because the hammer must drive approximately straight into the pin. If more angle is necessary for correct aim, a different mounting system must be used.

The pipe and firing assembly can be nailed to any suitable board, and the board then secured to something else. Or it can be wired or taped to things such as tree limbs, fences, furniture, and building blocks.

A pipe gun has the advantage of being able to shoot through objects such as curtains, windows, some doors, or walls. Two or more used in combination can cover a large area outside of a doorway or window. In that way a person who stands in what he considers to be a safe place and activates the trap can still be hit. Multiple targets can also be hit.

Anyone experimenting with this gun should be aware that it is potentially very dangerous to be anywhere close to it when it goes off. The hammer can be thrown straight back a good distance. The pipe can also burst and come loose from its mounting.

Molotov Cocktail

A molotov cocktail as described in Chapter 5 can be set on an arm to swing into a target and break. This can dis-

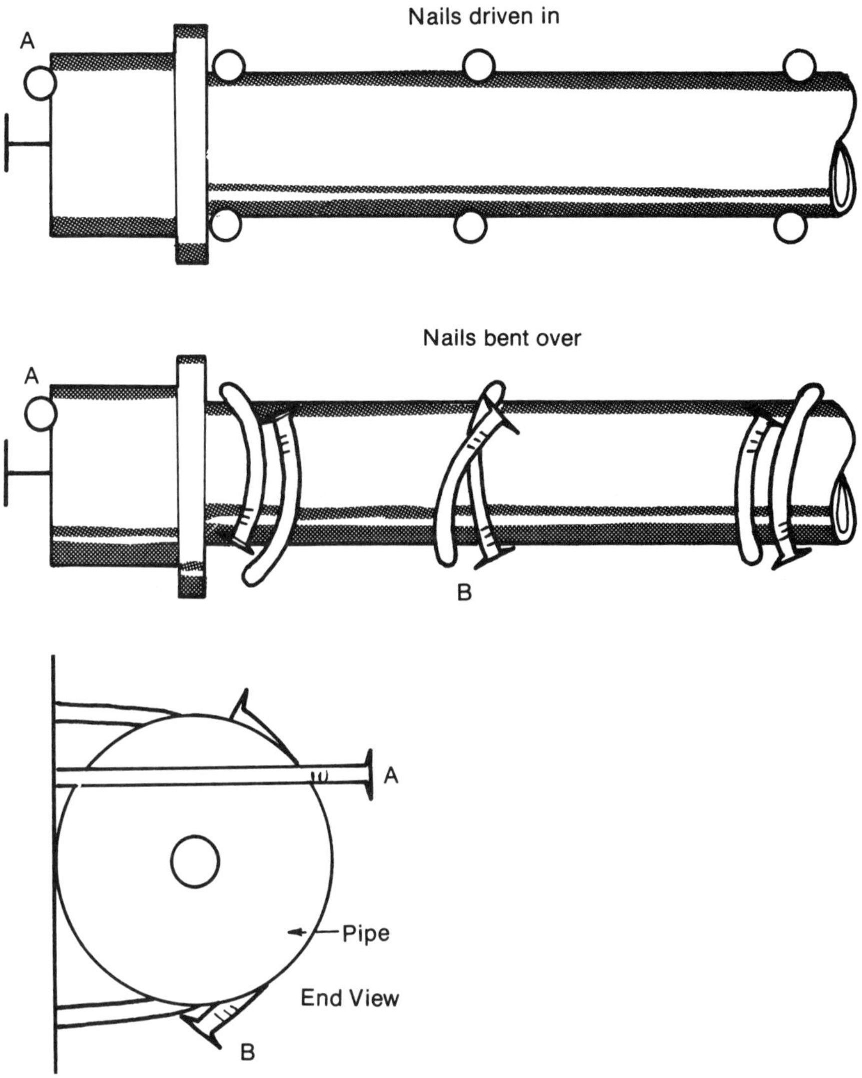

Fig. 55. Pipe gun mounting. The pipe lays flat against a wall or board. Nails are driven in at either side and one or two in back. The nails at either side are then bent over the pipe.

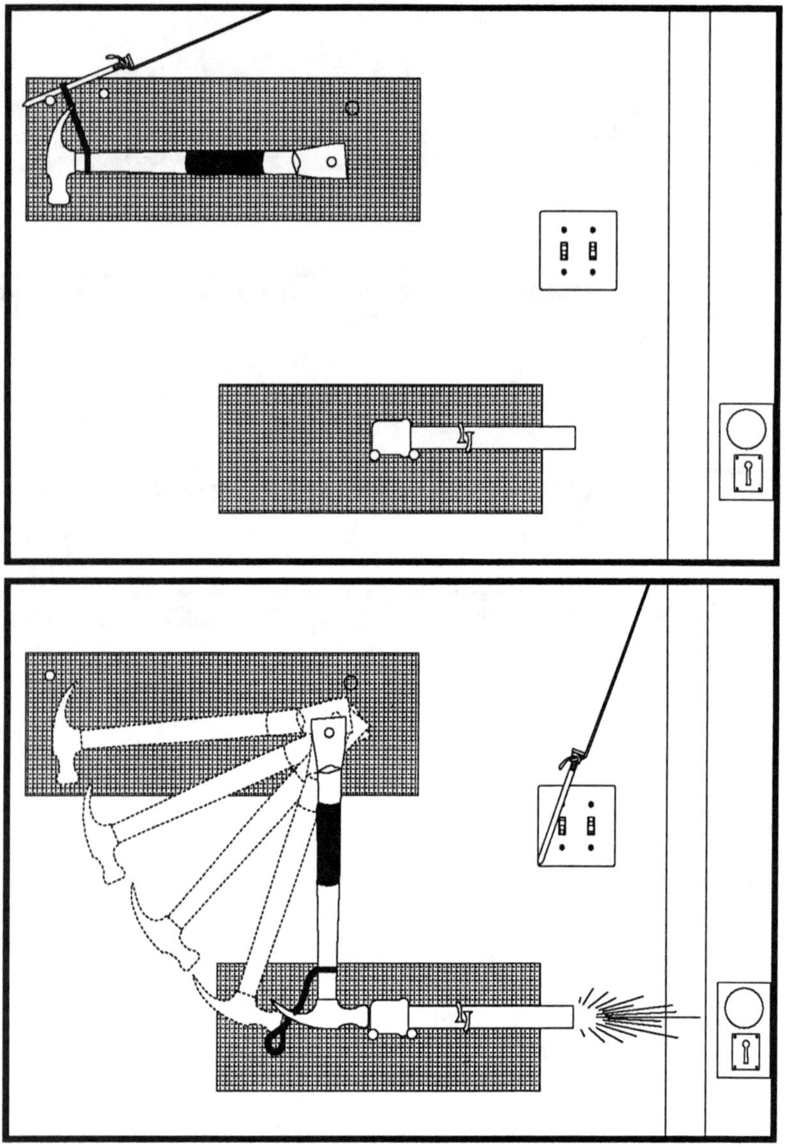

Twelve gauge pipe gun set to blow off a hand when the door is opened.

perse fuel better than one that breaks against a floor.

Two vessels with their arms hinged to the same anchor can collide in mid-air and disperse fuel very efficiently. They should be hung directly opposite each other (i.e. the two pulleys and one anchor form a straight line), so that they swing straight into each other.

Except for the arm, the mechanism is as described in Chapter 5. The igniter must be attached so as to allow the striker to be pulled straight out. Either one or both vessels can have an igniter.

Dumper

A dumper can be attached to an arm so that when the arm swings down it turns over and spills its contents. The arm can be hard or soft. The mechanism is the same as for the dumper described in Chapter 5, except that it swings as it dumps. This allows the contents to cover a much greater area.

With a hard arm, the vessel sits on and is atached to the arm. It can be glued, taped, tied, or attached in some other way.

A soft arm is attached at or near the bottom. It takes the place of the anchor line that tips a straight damper. The vessel will tip opposite the point it attaches to, and spill in that direction.

To keep the dumper from spilling all of its liquid before it completes its swing, the top can be partially covered. This will result in a better dispersal of liquid than if the top is left wide open. A small vessel can be partially covered by placing tape over the top. A lid with holes in it can be placed on any vessel. The amount of open area determines the rate of dispersion. The dumper will swing almost as far past the anchor as to it, and so can potentially cover a very large area, providing it does not dump all of its contents at once. Even if all of the contents are not dumped in the first full swing, they will be dumped as the pendulum swings back and forth.

Contaminating an Area

Contamination is anything undesirable that is in a place it is not supposed to be. The kind of contamination can make an area inaccessable, or make progress through it very slow or very dangerous. A device set to contaminate an area on triggering will allow an area to remain clean until invaded by an enemy.

The contaminant can be a chemical (i.e. nerve gas), radioactive material, or solid or liquid fuel.

Oil can be spilled on roads, hard floors, stairs, ladders, and other such areas. Oil is a very good contaminant, because it is difficult to clean up. Broken glass, tacks, nails, etc., on a floor will make an area inhospitable to a person not wearing shoes, or crawling. Large amounts of salt, chemicals (i.e. insecticides) and other poisons, decayed organic material, and so on can ruin a supply of drinking water.

Contamination can be delivered with a dumper, a breakable jar, or a trap door.

Welder/Pressurized Acetylene Bomb

Welders and pressurized acetylene bottles are commonly found together on construction sites, in metal shops, power plants, and other industrial areas. An energized welding electrode brought into contact with a grounded acetylene bottle should cause an explosion. This can easily be made to look like an accident.

The acetylene bottle should be as full as possible; this means both more fuel and more pressure. Fullness can be tested by barely opening the valve, and listening to how fast gas comes out.

The welder should be turned up as hot as possible, and the electrode (rod) should be a small diameter. It should be in good condition.

The bottle must be grounded. In most industrial areas welders are grounded to the structure, so that any steel in

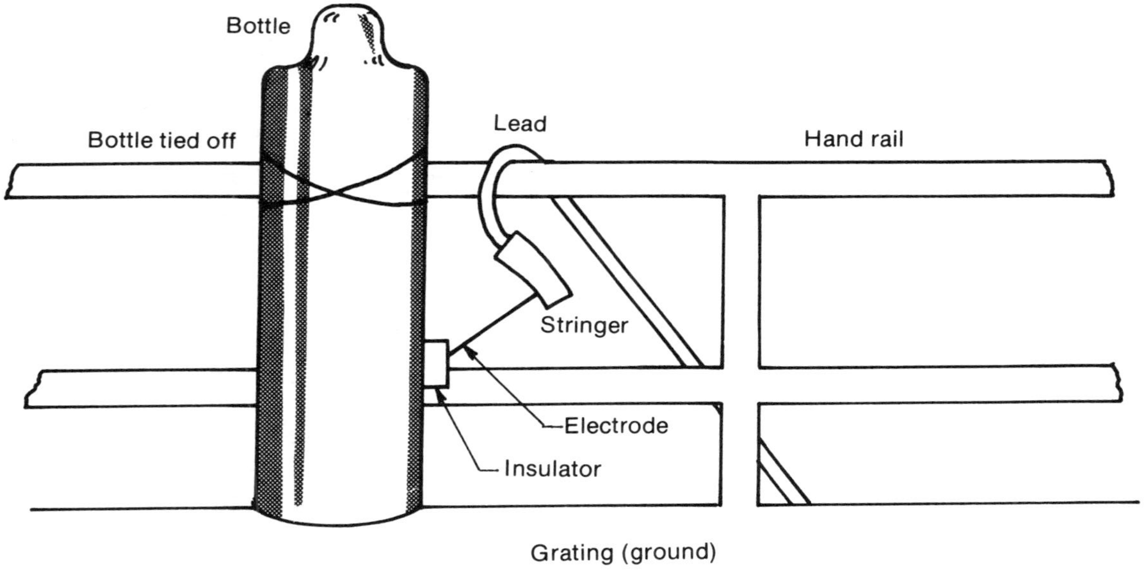

Fig. 56. Acetylene/welder bomb. The insulator is attached to the tripline. Since the surface of the acetylene bottle is curved, it is very important that the electrode drive straight in.

contact with the structure can be ground. This includes grating, hand rails, and support beams. It must be free of paint, rust, etc. A ground can be tested by being touched by the welder; if a flash is produced, the ground is good.

Acetylene bottles are usually stood against and tied to something such as a hand rail. A welding lead can be coiled over the same hand rail, with the stinger (electrode and holder) hanging down and electrode pointed toward the bottle. An insulator, such as a block of wood, or section of insulated lead is placed between the end of the electrode and the bottle, and attached to some sort of tripline, or made disturbance activated. When the insulator is removed, the stinger contacts the bottle. It quickly burns through, igniting the acetylene.

Since acetylene bottles are round, there is a chance that an electrode will slide off when it makes contact, and so not produce a fire. To avoid this, the electrode should drive straight into the bottle, as though attempting to touch its center. Fig. 56

Model Payloads

Building models of boobytraps will do a great deal to give a mechanic skill at building real ones. It is highly recommended that no attempt be made to build a real one without first building models. Since the nature of boobytrapping makes a mechanic something of a design engineer, even an experienced mechanic should sometimes use models before beginning construction. This is especially important for swinging deadfalls, because they are the most complex of the three forms.

One of the best model payloads is a bolt filled with washers. Rigging is done with kite string, and very small screw eyes or nails. Lashing is done with Super-glue.

Hanging from a long arm (i.e. ceiling to floor), the above payload can have a respectable amount of power. It can smash empty beer cans, potatoes, or bottles.

A claw hammer makes a good model hard arm. A match striker is a good model to make, because it requires good rigging.

Ceilings, the bottoms and sides of shelves, and similar places are good areas for model payloads to be built.

Practice Payload

A very good dummy payload can be made from a rubber ball. The load and anchor lines are attached by being tied around the center, and taped in place. When setting a payload of this type to hit a live target, the mechanic should be careful to avoid hitting the face and groin area.

Section III

Standing Deadfalls

CHAPTER 10

Form Structure and Characteristics

A standing deadfall is set over the target area with one part of the payload already touching the ground. The dead end (end touching ground) may or may not be attached to the ground, but either way it acts as a pivot point. The top falls in a curved path around the pivot point until it hits ground. The area under the entire length of the payload is covered. Since the payload falls downward, the target will be between it and the ground when hit. This can enhance power considerably.

If the dead end is not fixed to the ground in some way, the area under the high end of the arm is the most effectively covered. If the dead end is fixed, the area close to it is the most effectively covered, because the high end of the arm will pry down against that point, and so magnify the power of its weight. Fig. 57

This form of deadfall is very often used as an animal trap, because it is so easy to build: one end of a log is simply propped up. Its use against humans is limited, however, because it is relatively difficult to conceal. It uses its weight inefficiently, because it begins its descent with one point already touching ground. Given the same

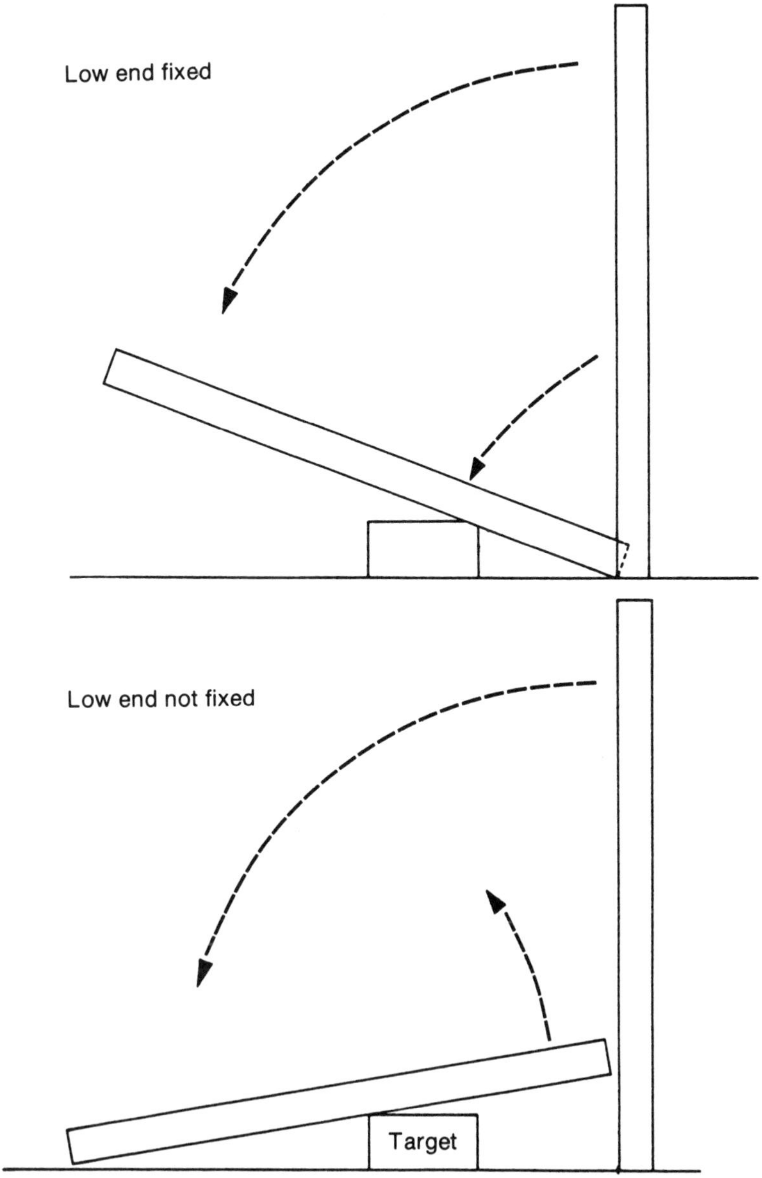

Fig. 57. How effectively the area close to the low end is covered is determined by whether or not it is secured against upward movement. Arrows represent force applied to the target.

amount of weight and falling distance, it will not fall as fast or hit as hard as either of the other two forms.

But it can still be very useful in some ways. Because one point remains in contact with the ground, rigging and triggering can be very fast and easy. Objects that are too large and heavy to be hoisted up and hung can be stood up and set to tip over. And, unlike the other two forms, it is not essential that a line run from a trigger to a point over the payload.

For the most part, this type of deadfall will be used when there is insufficient time or material to build one of the other types, and with payloads that are too large and heavy to be hung, but can be set upright.

CHAPTER 11

Triggering

A standing deadfall hung from a pulley will be triggered in the same way as the other two forms. If no pulley is used, the trigger will be something that tips the payload over. It will use a prop, balance, friction, or a combination of things.

No matter what type of trigger is used, triggering will be easier, faster, and more certain the higher the payload's center of gravity is. Center of gravity is raised by concentrating weight toward the top.

Center of Gravity

Every object has a point where all of its weight seems to be concentrated. This point is referred to as the center of gravity (CG). For the purpose of standing deadfalls, it can be said that gravity pulls down on the center of gravity, and ignores the rest of the object.

In order to stand, an object's center of gravity must be over its base. If it is not over the base, gravity pulls it down, and the object falls.

The higher the center of gravity is, the less it needs to move to be outside the base area. The same is true the smaller the base is. So an object with a high center of gravity and a small base is unstable, and will fall over easily. If the CG is closer to one side of the base, the object will be likely to fall in that direction.

If the CG is just barely outside of the base area, only a small part of the weight is being pulled down, and the payload can almost stand by itself. It can be supported by a load line, prop, etc., with very little effort. If the base is secured against backward movement, the support can be attached toward the low end. If the base is not secured, the attachment must be above the CG.

As center of gravity moves further outside of the base area, more effort is required to support the payload. Triggering becomes more difficult.

Prop Trigger

The most obvious way to support a standing deadfall is with a prop. A prop has the effect of extending the size of the base. When removed, the center of gravity is outside the base area, and the load falls. Since the payload CG is between it and the prop, falling is toward the prop.

Props are made sensitive by feel. The prop is placed so that it and the load are almost ready to fold. A prop can be set with its bottom almost ready to fall off a ledge, so that a small movement will cause it to fall. Some props can not be made sensitive, and something like a lever is required for a hair trigger. Props can take a variety of forms.

Simple Leaning Prop

A payload is stood leaning over, with a prop between one side and the ground holding it up. The degree of tilt of the payload determines how much pressure is on the prop, and so how easily it can be removed.

Often a prop will want to slide, either on the load or on the ground, either of which will cause it to fail. A notch or finger will hold the prop secure. Fig. 58

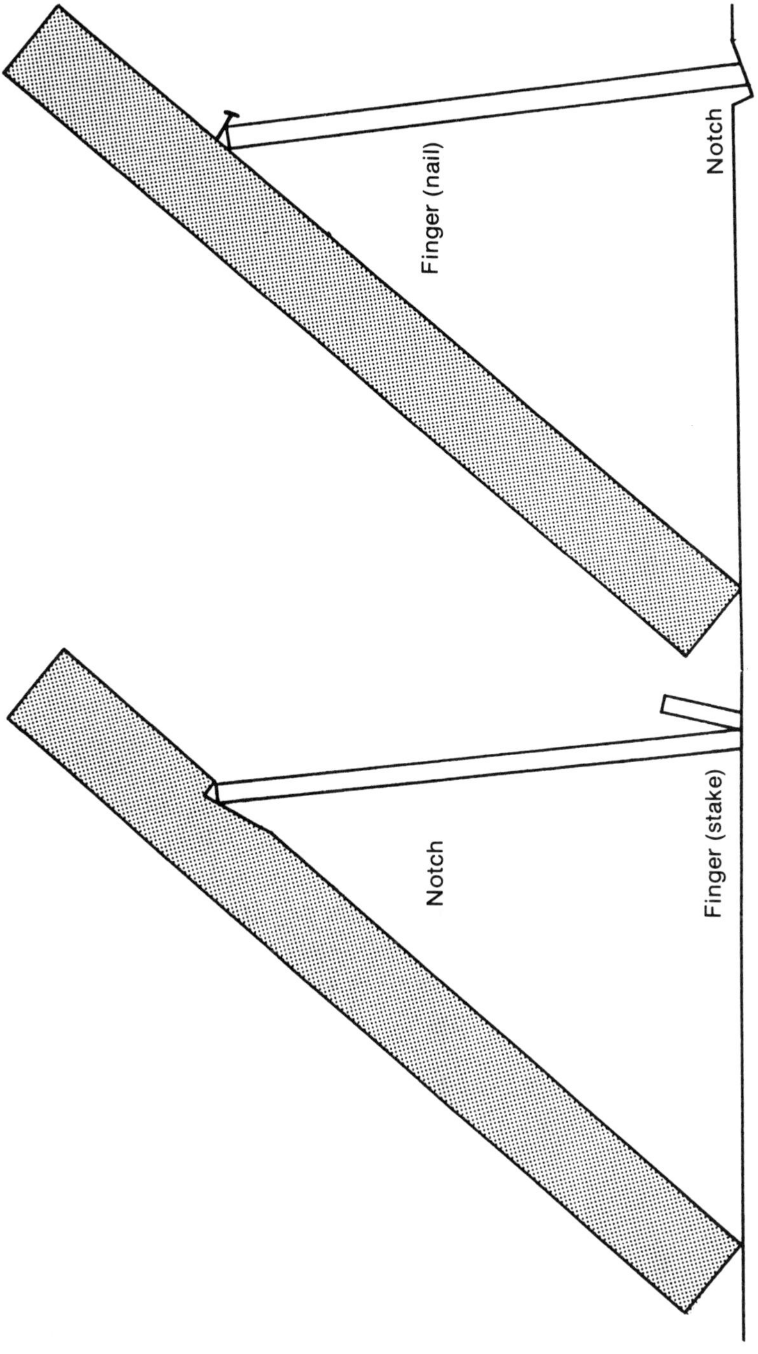

Fig. 58. It is often necessary to secure a prop to the load and/or ground to keep it from sliding. This is done by giving it something to catch on, either a notch or finger.

Since fall is towards the prop, the prop will to some extent occupy the targeted area. This can make concealment difficult or impossible. A prop can be disguised, or hidden by darkness. A prop that sits close to the base will be less in view.

Bottom Prop

One side of the payload sits up on a dead support, and the other on a prop. The two are of approximately the same height, so that the payload sits balanced and secure. When the prop is removed, the payload tips over. Since tipping is toward the prop, the prop is on the side toward the target. The prop and dead support must be high enough to cause enough tilt to be certain of a fall. If questionable, this can be checked by setting the payload between the dead support and ground, and seeing whether or not it will fall.

The prop should be structured so as to not interfere with the fall after triggering. If it does not pull well enough away from the payload base, it can affect the direction of fall by causing the low end of the base to move one way or the other. Depending on the nature of the payload, this may or may not be a problem.

If the payload is set so that its CG is just barely to the free side of the dead support, that support will carry most of its weight. That will make the load very stable, and also make the prop easy to remove, because it has less weight on it. A hair trigger can easily be set.

A trigger can be made more sensitive by dividing the prop in the middle. The tripline is attached close to the division point. The prop can be two or more food cans, two square-cut logs, bricks, and so on.

The prop should be as close to the free edge as possible, so that a minimum pull distance is required.

Prop and Roller

Any prop can be made a good deal more sensitive by installing a roller between it and either the payload or the surface. A roller can be set between two parts of a divided prop. Fig. 59

Door

One of the fastest and easiest ways to trigger a standing payload is to lean it against a closed door. The door is a prop, which will be removed by the target. When the door opens, the payload falls through the doorway. Concealment is no problem, because the door hides the entire trap. The entire trap mechanism can be made in just a few seconds.

A payload leaning against a door is unlikely to be as effective as one set with a more complex trigger. As it falls it is likely to be deflected by the door to some extent, and might miss the target altogether. In some instances the door might act as a shield between the target and payload. A heavy payload with a flat bottom will not be easily deflected by a door, and so is more likely to hit a target. Any payload will at least be a distraction.

The payload will fall in the approximate direction it leans.

Figure Four

A Figure Four is a sophisticated prop mechanism that can support a substantial amount of weight and still have a hair trigger. On triggering it folds as it collapses, and so does not interfere with the fall of the payload. It is complete in itself, with the trigger a part of the mechanism. The components are easy to get just about anywhere. For these reasons the figure four is the most commonly used and written about standing deadfall trigger.

It consists of three notched sticks. One is the actual prop, one is a lever that holds the deadfall suspended

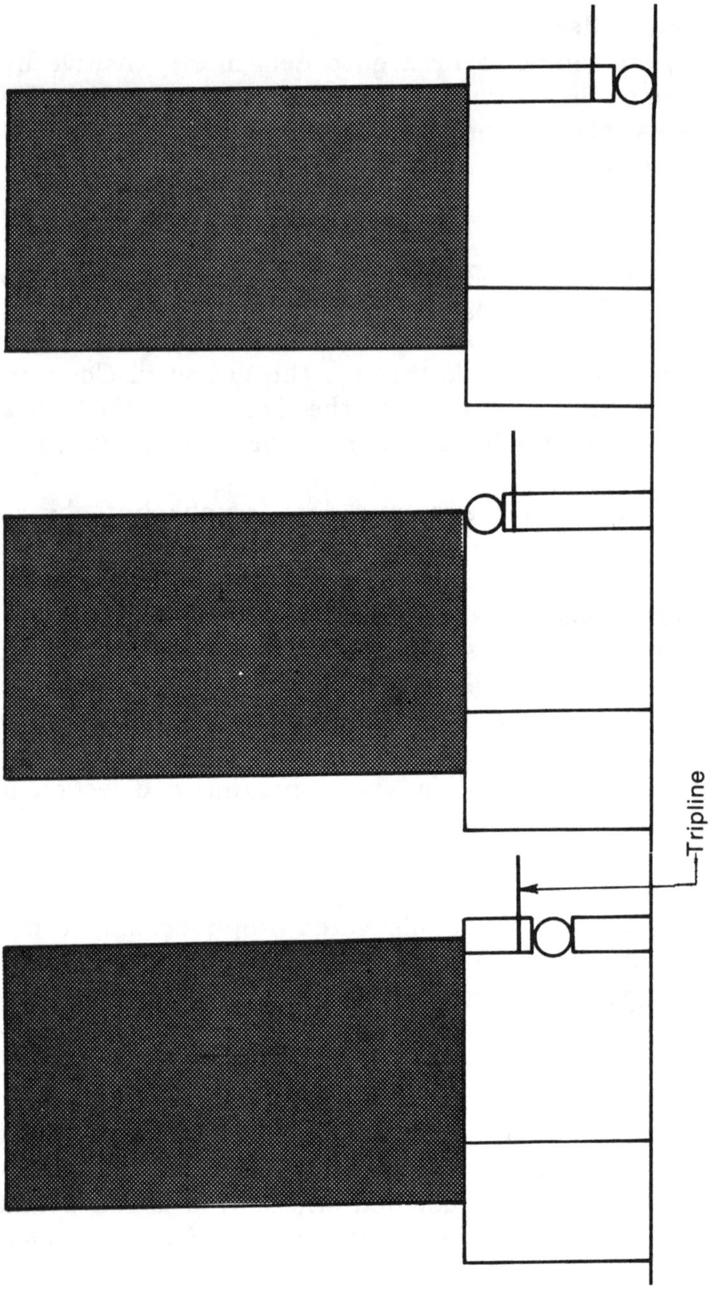

Fig. 59. Prop and Roller. A roller (sphere or cylinder) can make a prop very sensitive.

above the prop, and one holds the lever down.

The one that holds the lever down is the trigger stick; when activated, it releases the lever. The lever and deadfall together push the prop to one side and out of the way, and the load falls uninhibited.

The notches are what lock the sticks together. They work best if they are made deep and square. But they should not be so deep as to weaken the sticks, especially the lever stick. The prop stick is not notched at all. The lever stick is notched toward one end. That notch sets on top of the prop stick, and acts as a fulcrum. As a lever, it allows a small force to support a large weight. (It does not affect the amount of weight pushing down on the prop stick itself.) The more distance there is between the notch (fulcrum) and the far end of the stick (lever arm), the less weight the trigger stick has to hold. The trigger stick has two notches. One locks it to the lever, and the other locks it back to the prop stick, which is very stable because of the weight pushing straight down against it. Fig. 60

The trigger stick itself is of course another lever. The longer it is, the less force is required for release. Usually the figure four is set so that the trigger stick is approximately parallel to the payload, with the free end pointing toward the low end.

The mechanism is very strong, as long as it is directly below the weight and all of its components are in line with each other. When the trigger stick is moved, it either releases the lever, or pushes everything out of line. Either way, the mechanism collapses.

While this mechanism can be very useful for animal traps, its use with human traps is limited, because it can not easily be concealed.

The figure four can be modified by replacing the trigger stick with a cord held in place by a peg and lever assembly. Fig. 61

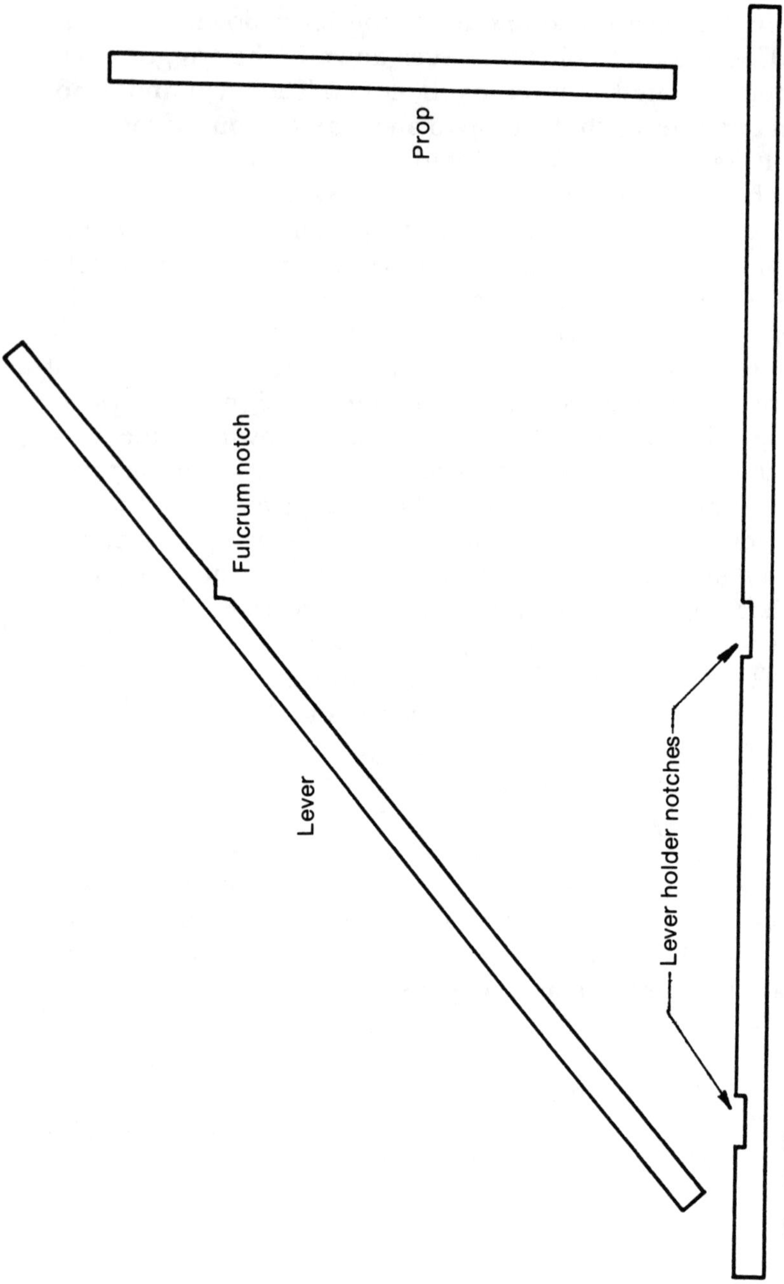

Fig. 60. Figure four trigger,

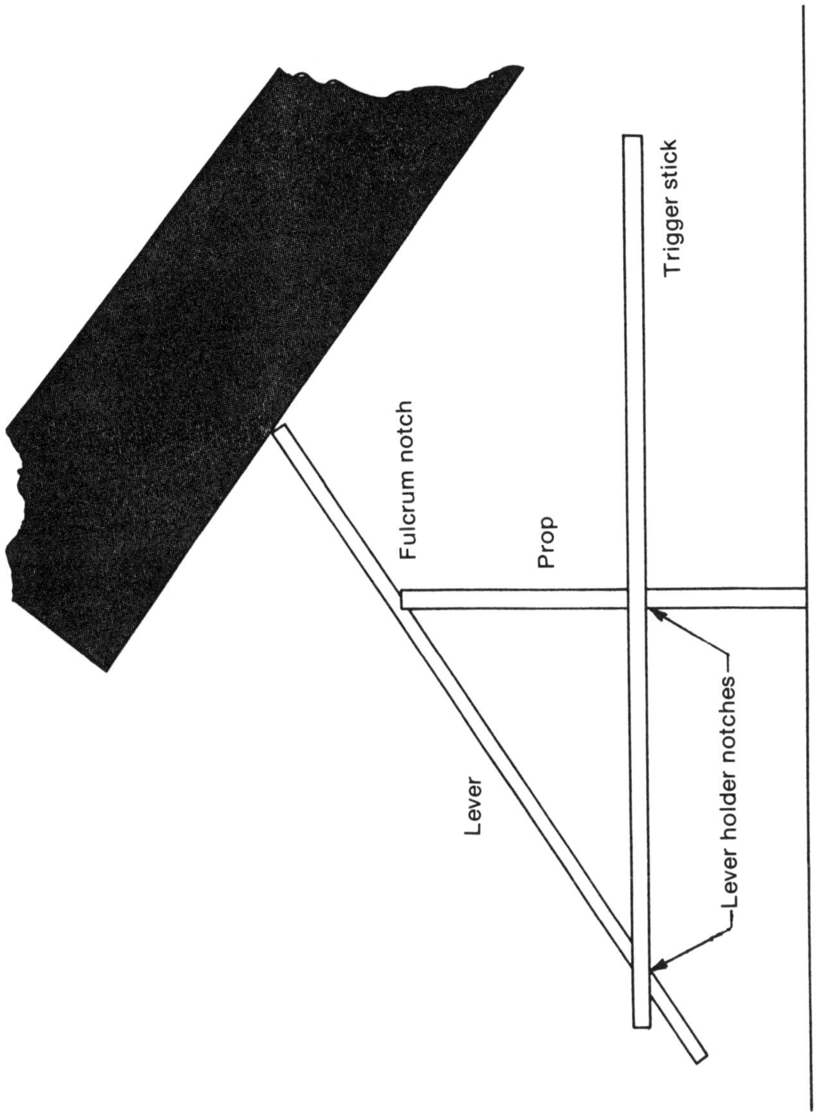
Trigger stick
Fulcrum notch
Prop
Lever
Lever holder notches

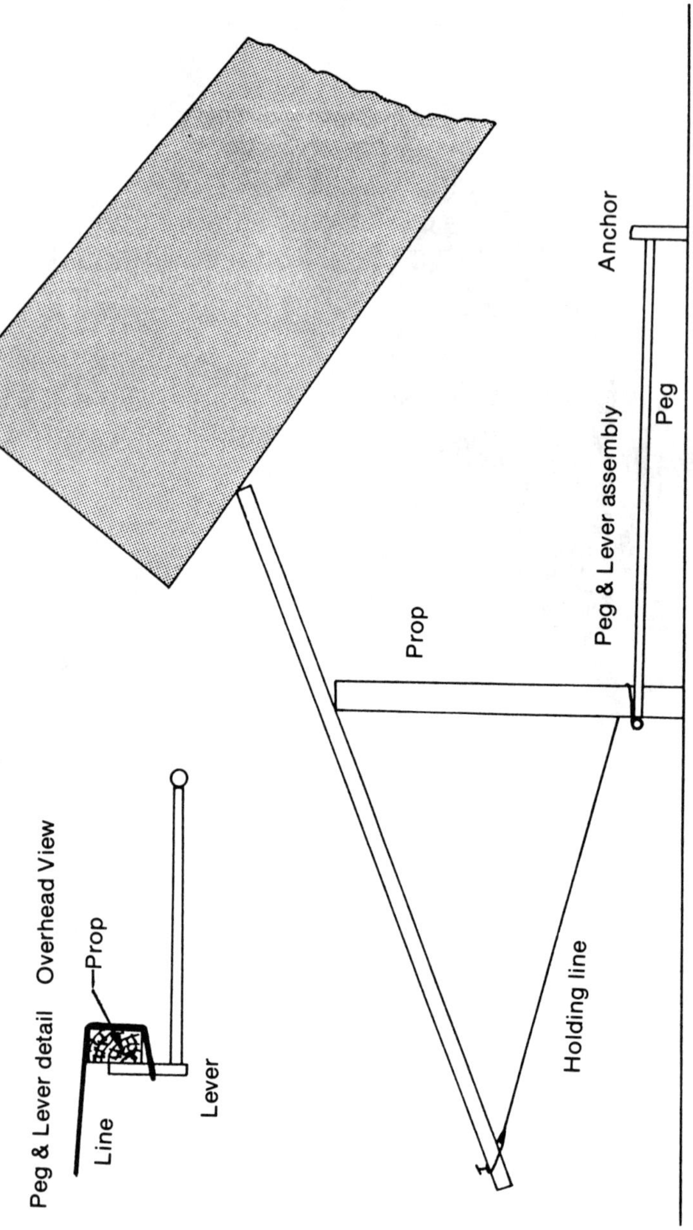

Fig. 61. **Modified figure four trigger. The trigger stick is replaced by a load line, anchor, and peg and lever assembly.**

Stacks

One of the most common causes of accidents is incorrectly-made stacks collapsing. Stacks can be built so as to fall on triggering, and direction of fall can be controlled.

A stack will stand as long as the center of gravity of each of its parts is supported. Center of gravity of each part varies as weight is set on it. As the CG of any one part gets closer to the edge of the member supporting it, the stack becomes less stable. When the CG of a member moves beyond its support, it and everything above it falls. The members below it may or may not also fall. The lower on the stack the first falling member is, the more of the stack will be certain to fall.

Fall of the first member will be in the direction its CG pulls it. Each part will be controlled to some extent by its own CG, along with the direction and amount of force applied to it from above and below. Some stacks will fall in a single direction, as though they were all one piece. Sometimes top members will push lower members back in the opposite direction they fall in, causing the stack to fall in two directions. A stack made with the CG of its parts arranged toward opposite sides of the base can fall in every direction, or collapse around its own center.

Only one part of a stack needs to be attached to a trigger, and only one part needs to be offset. The trigger may or may not need to be attached to the offset member. The amount of offset determines trigger sensitivity.

The effectiveness of a stack can be improved by making the top larger than the bottom.

A stack can be activated the same way as any other standing deadfall. This includes with a tripline, lever, or treadle. A stack can be made so sensitive as to be disturbance activated.

A good example of a disturbance activated stack is cement blocks (counter weights) stacked together on a railroad car. When another car makes contact with it, or whenever anyone climbs on it, it falls.

Balance

A payload can be stood balanced upright, with a trigger attached that will tilt it enough to tip it over. If the payload is almost ready to fall when the trap is set, it will have a hair trigger. If it is too much off balance, it might trip prematurely.

A tripline can be tied directly to the payload to pull it over. A lever can push one over.

A payload can be set on a board sitting on a fulcrum (treadle) that will tip the payload over when stepped on. A treadle can be easier to make and hide if a shallow pit is dug under the part that the target steps on. The edge of the pit acts as a fulcrum. A deeper pit can make a trap a good deal more deadly.

With a treadle or lever, fall can be toward or away from the target. Away from the target is usually easier, and can be made much more quickly. A trigger that tips a payload away from the target can be used with noise makers, fire traps, etc.

Force on a treadle can tip a payload above or below a target. This is done by installing a spacer between the treadle and payload. Fig. 62

As with all treadles, trigger sensitivity can be increased by adding weight to the treadle. This will be necessary with a heavy payload that has to be lifted to be tilted.

Friction

A payload can be stood leaning against a support (i.e. wall, tree limb) almost ready to fall, with friction between the payload and supporting object counteracting the force of gravity. This is used together with balance, and props.

Timing

Falling rate is variable with standing deadfalls. If the payload is just barely off balance when it begins its descent, the fall will begin slowly. The further outside the

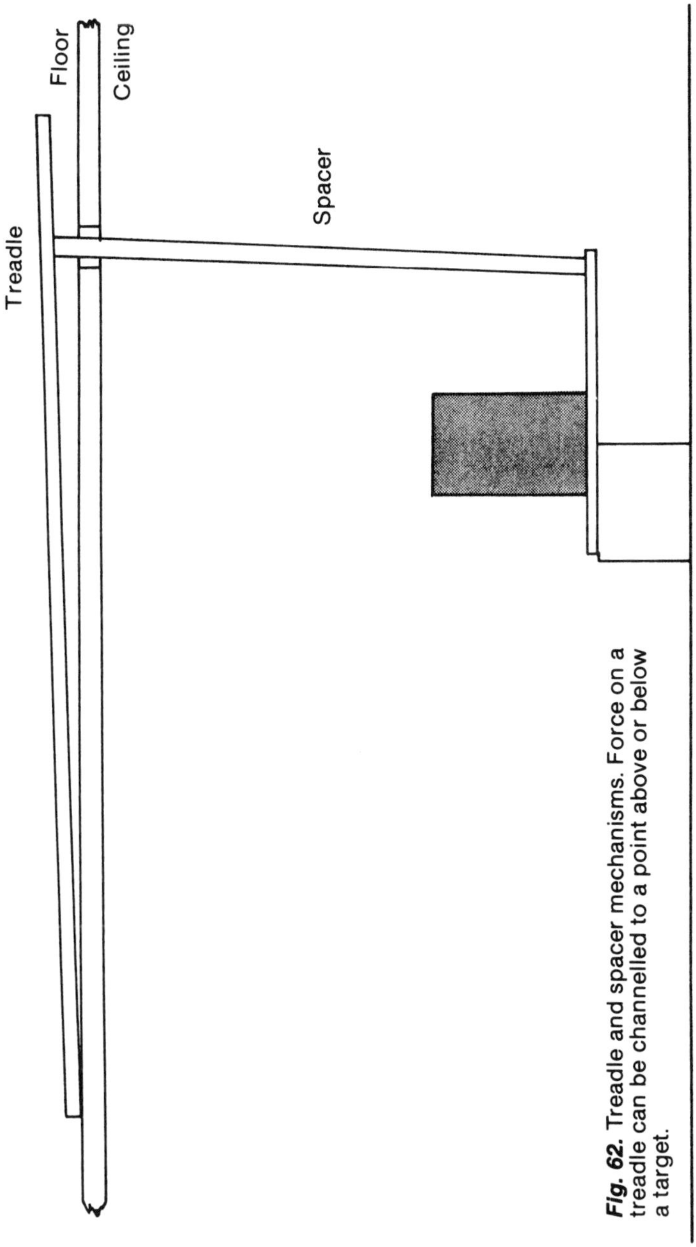

Fig. 62. Treadle and spacer mechanisms. Force on a treadle can be channelled to a point above or below a target.

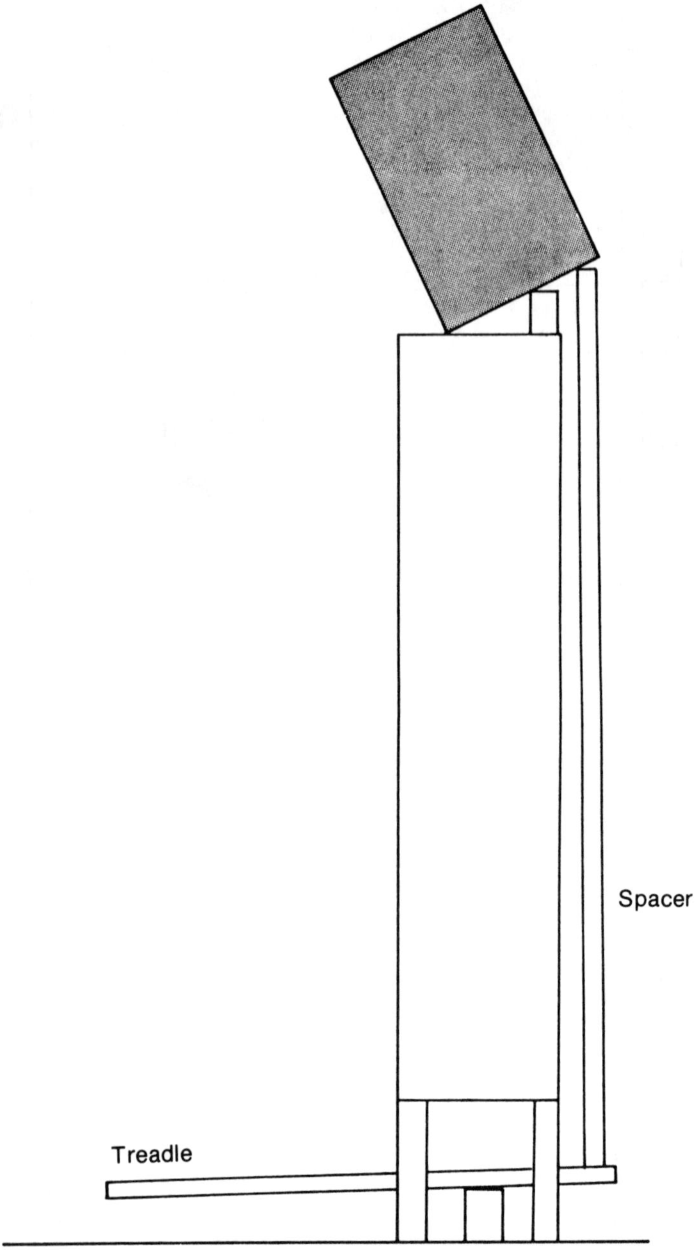
Spacer
Treadle

base area the CG is when the fall begins, the faster the fall will be.

In every other respect, timing is essentially the same as with straight deadfalls. Fall rate will be the same, slightly slower, or much slower, depending on degree of tilt. If questionable, fall time can be determined by letting a payload (or an object of similar size and weight) fall from the position it will be set in, and timing the fall.

Timing is simplified by setting the payload so as to keep the target in the kill zone as long as possible. This is done by setting it to fall approximately parallel to the target's path.

The relative slowness of standing deadfalls can be compensated for by setting them to fall in the same direction the target is traveling.

CHAPTER 12

Rigging

The biggest advantage of standing deadfalls is that rigging can be very fast and easy. Unlike the other two forms, this one does not require a load line, pulley, and so on. Even very large and heavy payloads can be set with relatively little effort, because the payload stands rather than hangs; one point remains in contact with the ground.

A standing payload can be held upright by a load line and pulley, in which case rigging rules as covered in Chapter 3 apply. Otherwise, rigging involves simply standing the payload up, and causing it to fall toward the target.

A standing deadfall can be improved by attaching the low end to the ground. In this way the arm can act as a lever, with the top prying down against any part of the arm that contacts the target. Given the right circumstances, this can magnify power considerably, as well as increase the size of the area effectively covered by the payload.

It must be attached with a hinge that will not interfere with the fall. This is easy to arrive at, because the hinge only needs to stop upward movement. It can be done with a U-shaped saddle, or by tying a rope between the end and an anchor. Some slop in the hinge will not hurt anything.

Weight concentrated at the bottom can act as a hinge. A barbell with weight concentrated at both ends is a good example of weight distributed to perform two functions: hinge and pry.

In some cases it is better to leave the low end unattached, because some payloads will stand up and fall again (roll over), so covering more area. Some will bounce. A payload set at the top of a staircase will cover the entire staircase if it is left unattached.

Since the target is between the payload and ground, the angle of attack has an effect on the amount of damage done. The straighter the payload drives into the ground, the more force the target must absorb. A payload that attacks from the side will push the target; one that attacks from overhead will smash it between two surfaces, just like a straight deadfall.

The taller the payload is, and the further from the pivot point the target is, the straighter into the ground the payload will be driving when it hits the target. Raising the pivot point (by setting the payload up on something) will have the same effect.

As with all deadfalls, falling energy is a product of weight and falling distance. Increasing either quantity will increase power. A standing payload can be thought of as an upside-down swinging hard arm; certain principles of power apply equally well to either form. This is especially true when the low end of the payload is fixed.

Concentrating weight all the way to the free end makes maximum use of the available weight by allowing it to fall as far as possible. As long as the low end is attached to the ground, weight at the top will apply pressure to any point on the arm, and the pressure will be magnified by leverage.

The higher the top of the payload is when it begins its

fall, the further it will fall. So falling distance is determined both by payload length and degree of tilt. The taller it is and the straighter it stands, the more power it will have.

Falling distance can also be increased by setting the payload up on something. This can be a table, tree limb, ledge, or anything else that will hold it. A payload standing at the top of a flight of stairs, ladder, cliff, etc., will have more power than one that is on the same level as the target.

Aim

The payload will fall in the direction of tilt, and so is aimed by leaning toward the target area when set. It will cover an area more or less equal to its size. The top falls in a curved path around the pivot point (low end), until it hits ground.

A falling payload can be deflected by hitting other objects on the way down. Things like tree limbs used to hide the payload might make it miss the target.

An easy way to get a payload to fall straight is to set it against something like a wall or limb so that it slides along it as it falls. A flat-bottomed payload will not be deflected easily. Guides can be placed along the bottom to hold it on course.

CHAPTER 13

Payloads

A payload will usually be an object used more or less as is. Some straight deadfall mechanisms can be adapted and used with standing payloads, and some hard arms can be set upside down as standing deadfalls. But usually if a payload is built it will be one of the other two forms. Most of this type will be quickly and easily improvisable.

Trees

A standing tree can be cut and set to fall quickly and with relatively little effort. It does not need to be hoisted, because it is standing upright to begin with. It is easily concealed, because it looks like a tree standing where a tree is supposed to be. Large trees can do a great deal of damage when they fall. One tree can be set to trigger more trees to fall, and so cover a large area.

A falling tree can stop or wreck a vehicle, cave in a structure (i.e. house, tent), and attack multiple targets. One tree can fall across a road to stop one or more vehicles, and a second one can fall on the road to smash them. The first tree can pull the second one down behind it.

Just as most of the traps covered so far require some

mechanical skill to be made right, this one requires skill at felling trees. An experienced woodsman who has felled many trees can cut and set one or more to fall quickly and easily, and the trap will have a good chance of success.

Tall straight trees with insubstantial branches, such as many pine trees, work best. A tree must be able to fall without its branches or those of another tree inhibiting or stopping its fall. Large branches can be trimmed off; small ones will have no significant effect. Very large branches can effect a tree's balance, and so alter its fall.

A tree that leans will want to fall in the direction it is leaning in. If a selected tree leans in a direction away from the proposed target area, some lean can be taken out by cutting a notch and hammering in a wedge. This takes time and skill, and only so much lean can be taken out. So, if possible, a tree that does not lean, or one that leans towards the target area, should be used.

Wind can have a significant effect on direction and time of fall. A tree set to fall on triggering can be knocked down prematurely by a strong wind. If a tree is triggered when the wind is strong, wind can alter its course. The amount to which wind affects a tree's fall depends on the amount of surface area the tree has for the wind to blow against, weight of the tree trunk, and the force of the wind. The less branches a tree has, the less surface area it has. A tree surrounded by other trees will be less affected by the wind, because the surrounding trees can block much of it.

To set a tree to fall, two notches are cut in its base. The first notch is cut on the side that the tree will fall towards, and extends halfway through the tree. The bottom of the notch is cut flat, and the top is cut at about a forty-five degree angle. The second notch is directly opposite the first, but two or three inches higher. Depth of the second notch is variable. It should be cut carefully, until the tree begins to show signs of being ready to fall (swaying, cracking

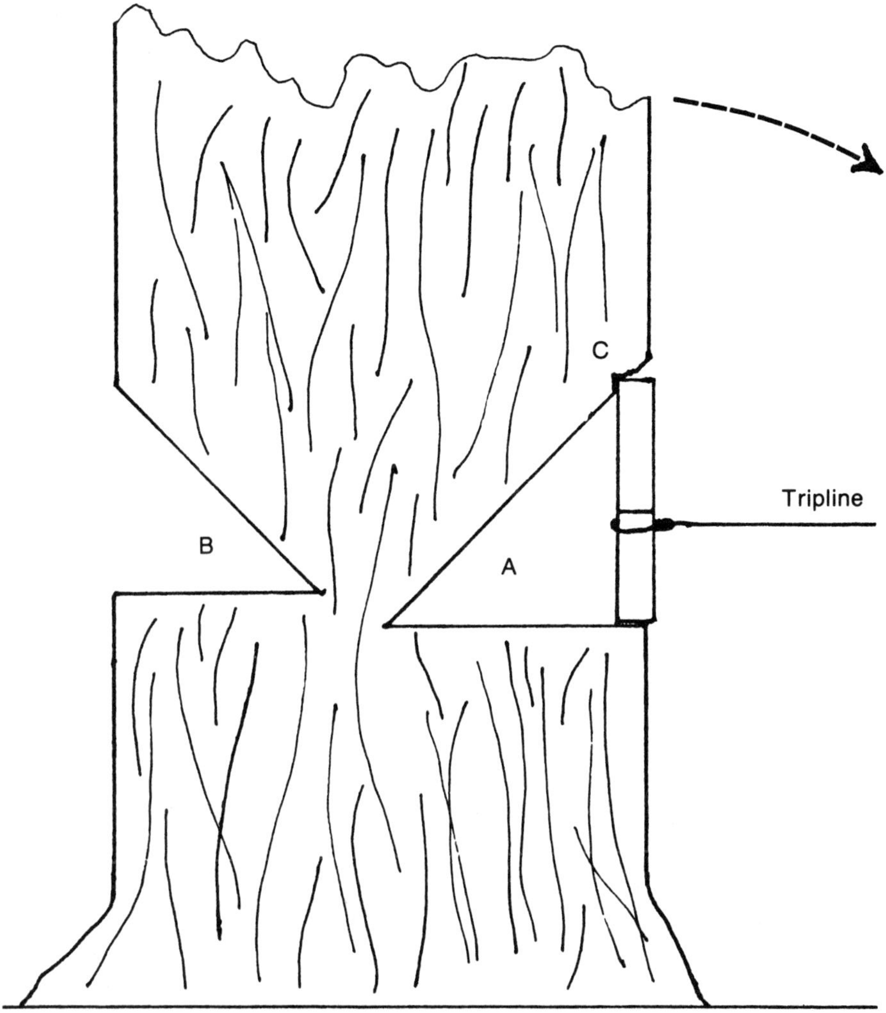

Fig. 63. Treefall. The tree will fall toward the lower notch (a). A divided prop holds it up. Notch (c) holds the prop secure.

sounds). At that point the notch is deep enough. Or it can be made deeper, to put more pressure on the trigger. Fig. 63

Triggering can be done with a prop, load line and pulley, or in some other way. The trigger mechanism should be installed before the second notch is cut, so that the tree can slowly put pressure on it.

Telephone Poles

Telephone and power line poles can be cut and triggered in the same way as trees. The lines they support can keep one from falling, or effect the course of the fall. This will be less of a problem if several poles next to each other are all cut. Only one needs to be attached to a trigger; it will pull the rest down behind it.

Road Bar

Either of the previous two payloads can bar a road. Anything long enough to go across a road and high enough to stop a vehicle can be a road bar. A bracket set opposite the pivot end can catch and hold the falling end, and hold the bar at a specific height. Fig. 64

Door Slammer

There are several ways that a standing weight can push a door shut as it falls. It must push in the right direction; one that stands close to the door frame should fall approximately parallel to the frame.

A heavy door can deflect a light payload, which might slide along it and fall free instead of slamming the door. A heavy payload with a flat bottom will be unlikely to be deflected. This can be a refrigerator, chest of drawers filled with weight, one or more railroad ties, and so on.

A heavy enough payload can effectively bar a door after slamming it. Fig. 65

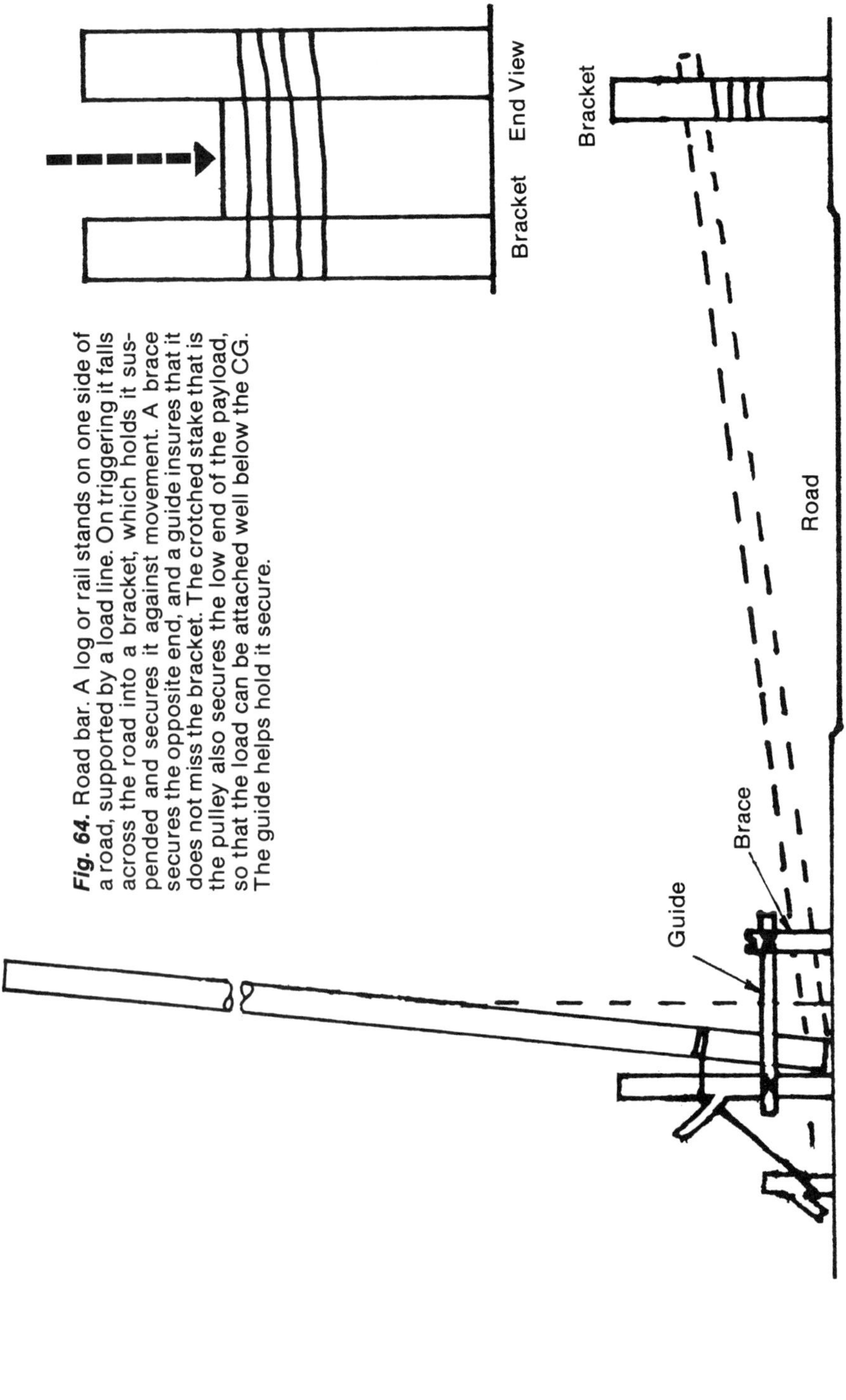

Fig. 64. Road bar. A log or rail stands on one side of a road, supported by a load line. On triggering it falls across the road into a bracket, which holds it suspended and secures it against movement. A brace secures the opposite end, and a guide insures that it does not miss the bracket. The crotched stake that is the pulley also secures the low end of the payload, so that the load can be attached well below the CG. The guide helps hold it secure.

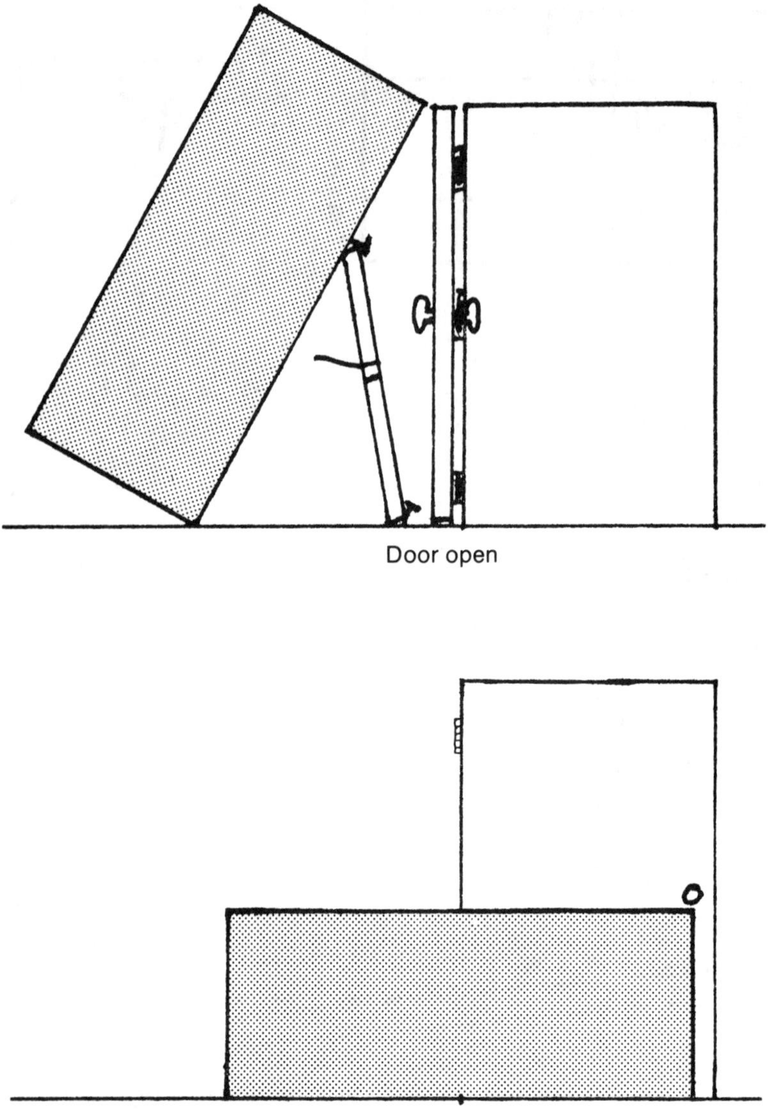

Door open

Door closed

Fig. 65. Door slammer. A large, heavy, flat bottomed payload pushes a door shut, and locks it.

Barbells

A set of barbells makes a very good standing payload. If all the weight is concentrated at the top, and the bottom is hinged in some way, it can be very powerful, especially if it sits up on something. Knives or other spikes can be attached to one or more of the weights (see Chapter 9).

Barbells in their normal state have weight evenly divided between the two ends. If stood in this way they will have less power than if all the weight is at the top, but they can still be very effective. Weight at the bottom holds the bottom down, and so acts as a hinge, making lashing unnecessary. Using them as is allows a trap to be set very quickly. Fig. 66

Miscellaneous Weight

Anything large and heavy enough to inflict damage can be stood and triggered to fall. The more simple and symmetric the shape is, the easier it is to work with.

Payloads of this type can be refrigerators, book cases, chests of drawers, sofas, lamps, logs, railroad ties and rails, boulders, and so on.

Dumper

The easiest way to cause a vessel to spill its contents is to tip it over. All of the dumpers covered so far can be set standing instead of hanging. These will usually be easier and faster to make.

Rather than dumping fuel on the target from above, standing dumpers will spill fuel below the target. Since fire burns up and smoke rises, they can be very effective used in this way.

Vessels can be leaned against doors to fall over and spill when the door is opened. A glass can be set on a shelf in a cupboard to spill when the cupboard opens. Other doors can be rigged just as easily. Igniters can be installed as explained in Chapter 5.

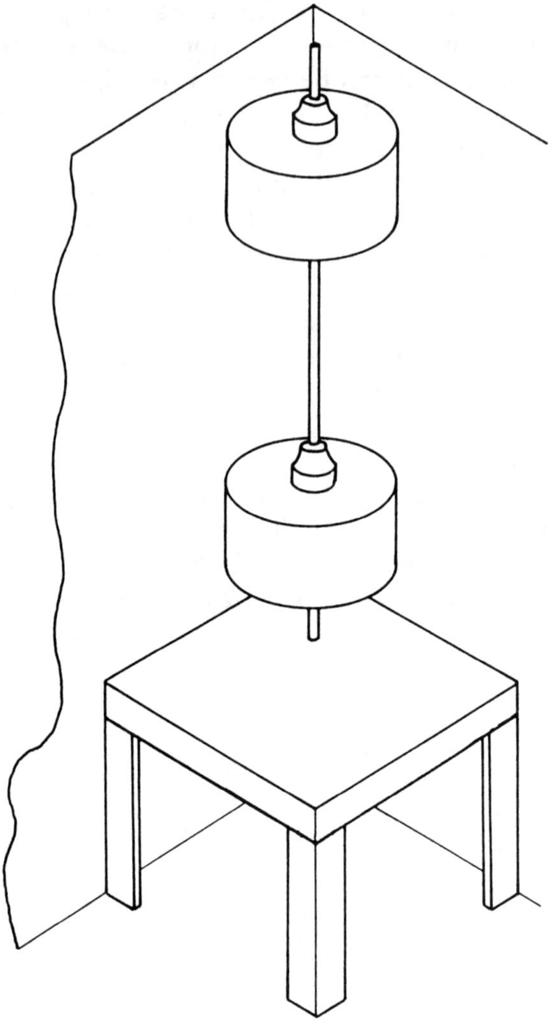

Fig. 66. Barbells used in their natural state require very little time to set. The bottom weights act as a hinge. Something like a long coat can be thrown over the top so that it looks like the coat is hanging on the wall, and the barbells are hidden. When the target pulls on the coat, the payload falls, in the direction it is being pulled.

There are many ways dumpers can be set to tip over. They are especially well-suited to traps set in a basement, with solid fuel (i.e. paper, wood) piled around them. Used in this way a building can quickly be turned into an inferno. Fire should originate from escape routes, and the trigger should be deep inside the structure.

Burners: Candles, Lamps, and Torches

Back in the days when artificial light came from some sort of burner, accidental fires were commonly started by lights being tipped over. Candles, lamps, and torches are all sources of fire, which can start bigger fires. Lamps contain reservoirs of liquid fuel, which makes them potentially very dangerous.

The first lights used by man other than camp fires were wood splinters set in a holder at an angle that allowed them to burn slowly and continuously. Later the splinters were saturated with better fuels such as pitch, which allowed them to burn a great deal longer and at a different angle. They were improved again by one end being wrapped with absorbent material which would allow them to hold more fuel, and so burn longer and brighter, and be held more easily. This last improvement ultimately became a torch.

It was eventually learned that containing the fuel would prolong burn time a great deal by limiting the amount of fuel exposed to air. Ultimately the absorbent material became a wick, and the container a lamp. The handle was no longer necessary.

Candles, which could burn for a very long time, also came into use. They were made by encasing a burnable wick with solid fuel, such as tallow (fat).

Any of the above devices can be set to fall over and ignite a fuel supply. Or fuel can be dumped on them. It should be remembered that fumes from a supply of liquid fuel can be ignited prematurely by an open flame. This danger is minimized by keeping the fuel vessel covered.

Lamps

Lamps contain the same elements as a molotov cocktail: liquid fuel contained in a vessel, fire, and usually a wick. The first lamps were made by filling clay or stone bowls with fuel such as pitch or tallow, and lighting it. Since fuel has to be in contact with air to burn, only the fuel on top would burn, with the rest waiting in reserve. Eventually wicks were used to suck fuel from a reservoir to a flame. This allowed a slower burn rate, and a cleaner (less smoke) flame. In order to work right, the wick had to be kept away from the fuel reservoir in some way. At first this was done simply by laying the wick over the side of the lamp. Later it was held by using teapot-shaped vessels, with the spout holding the wick. Since then many more improvements have been added.

Just as primitive lamps can tip over and start fires accidently, forms of primitive lamps can be made using the same principles, and set to spill burning fuel on triggering. Any liquid fuel can be used. Thick fuel (oil or gasoline mixed with oil) burns longer and has better effect when it spills. A wick is not mandatory, but is a substantial improvement. Any absorbent material, such as cotton, moss and so forth can be a wick. The material should be white (undyed or bleached) if possible. Anything that can contain fuel can be a lamp.

The biggest disadvantage of lamps is that from the time they are set they are consuming fuel. So the more time there is between setting and triggering, the less power they will have. This disadvantage is offset by the fact that they require little time or effort to make. Their best use is when it can be known that an area will be occupied by an enemy soon after the trap is set, and there is insufficient time or material to make a more complex device.

Simple Lamp

An open vessel is filled with liquid fuel, and set to spill

on triggering. The surface of the fuel is then lit. When it spills, burning fuel will pour out. It can spill directly on a target, or on a larger fuel supply. The vessel can be a bowl, cup, bucket, or drum. Burn rate can be slowed by covering part of the vessel.

This device can be very dangerous, and it is highly recommended that the mechanic leave the area immediately after lighting the fuel.

Lamp with Wick

A jar with a hole punched in the lid or a small-mouthed bottle is filled with liquid fuel, and a rag is stuffed into it, extending to the bottom, and with one or two inches sticking out the top. The rag is saturated by tipping the bottle over. The bottle is then set to fall over and either break or spill on triggering, and the wick is lit.

If set to spill, fuel will pour slowly out the mouth (depending on how tightly stuffed-in the rag is), and will be ignited as it reaches air. This can then ignite more fuel.

If set to break, it must be set over a hard surface, and allowed to fall far enough to break. It can be set-up on something such as a stack of books, that will fall over. Several fuel jars can be stacked on top of each other, with only the top one burning. A bottle can be taped to something like a broom handle, and leaned against a wall or door.

A wicked lamp can burn for one or more hours, providing the wick extends to the bottom of the reservoir.

Candles

Candles have the advantage of a very long burn time, and the disadvantage of a small, easily extinguished flame. Liquid fuel poured directly on a flame will very likely put the candle out and not ignite. A candle set to tip over might blow out on the way down. Even if it stays burning, the small flame can not be expected to start a good fire, unless properly nurtured.

As mentioned in Chapter 5, a tent built over a candle will protect it from liquid fuel, and still allow fumes to get through and reach the flame. The tent is made out of cotton or other absorbent material. It can be a miniature model of any style of improvised shelter designed to protect people from rain. It must be high enough over the candle to allow a good flame, and not be burned. It is very important that it allows sufficient oxygen to reach the flame, both before and after ignition.

A candle set to tip over should fall no farther than necessary. The landing should be soft. A cushion of soft tinder will allow a soft landing, and bring the flame into contact with fuel that can be ignited easily by a small flame.

Tinder can be very fine slivers of wood beaten until soft, or shredded inner bark of some trees. Shredded and crumpled paper, especially tissue paper, can be very good. Gunpowder is excellent tinder. No matter what material is used for tinder, it should be very dry, and it should allow for a very good air-fuel mix. It should be tested for flammability before use.

Sprayers

An aerosol container can be activated by a weight set on the button. Some can be activated by a weight shearing off the button. The contents can be paint, fuel, or anything caustic.

Two of the best fuels that come in spray cans are ether, which is sold as engine starting fluid, and thin oil, such as WD-40. Hair spray is good. Anything marked **Danger: Extremely Flammable** will work well.

Spray fuels can be used in combination with igniters in a variety of ways. The spray and igniter must both be aimed so as to contact each other.

The best caustic liquid in a spray can is oven cleaner. Anything marked **Keep Out Of Eyes** is good. Caustic sprays should be aimed at the face, especially the eyes.

Anything that has to be mixed (shaken) prior to use is of limited value. This includes most paints.

The weight should be in contact with the button when the trap is set, so that no actual falling takes place. It must be set so as to remain in position and hold the button down. The can must be secured in position.

Electrical Switch

There are several ways that a circuit can be completed by a standing deadfall mechanism. This switch is disturbance activated.

Several inches of insulation are stripped from the end of one lead, and a circle is formed from the bare wire. A long slender conductor (i.e. steel or copper rod, pipe, nail, bolt) is stood in the center of the circle, with the bare end of the other lead wrapped around its base. The circuit is completed by the standing conductor falling over and contacting the circle of bare wire. Fig. 67

Trigger sensitivity is determined by how easily the conductor falls over.

If something like a bolt or nail is used, it must be made certain that the head does not prevent contact. This can be done by raising the wire circle above the platform. Fig. 68

Walls

A wall is a standing trap door. Things such as logs, rocks, fuel, and so on can be set behind a wall and triggered to fall over. Power is increased if the wall is high overhead, as in mountain regions.

A brick wall is a stack of bricks. If the bricks are loose, the wall can be triggered to fall the same as any other stack.

Since walls hold ceilings up, one or more walls falling can drop a roof. Structures can be built to collapse on themselves. Walls can be set to fall inward or outward, or collapse around their own center.

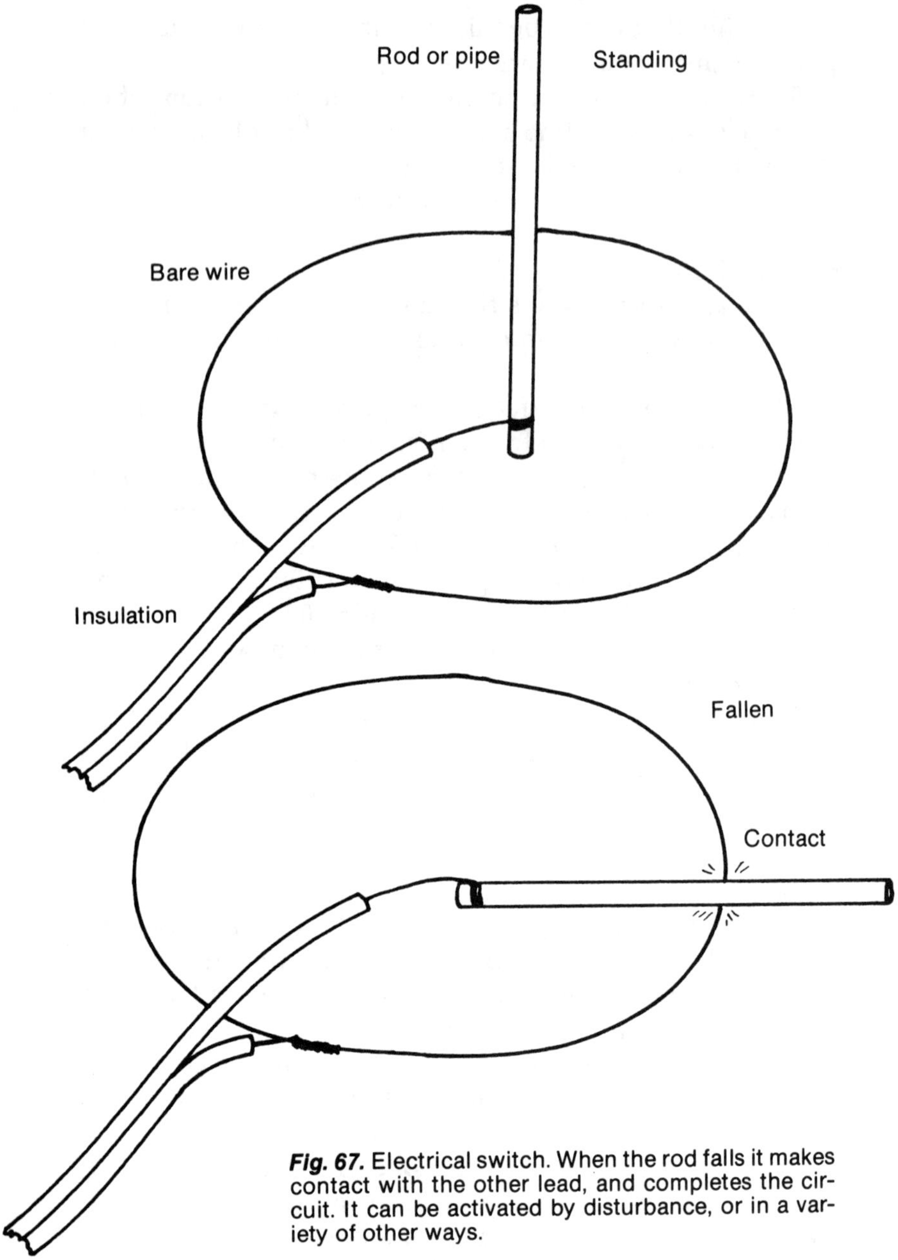

Fig. 67. Electrical switch. When the rod falls it makes contact with the other lead, and completes the circuit. It can be activated by disturbance, or in a variety of other ways.

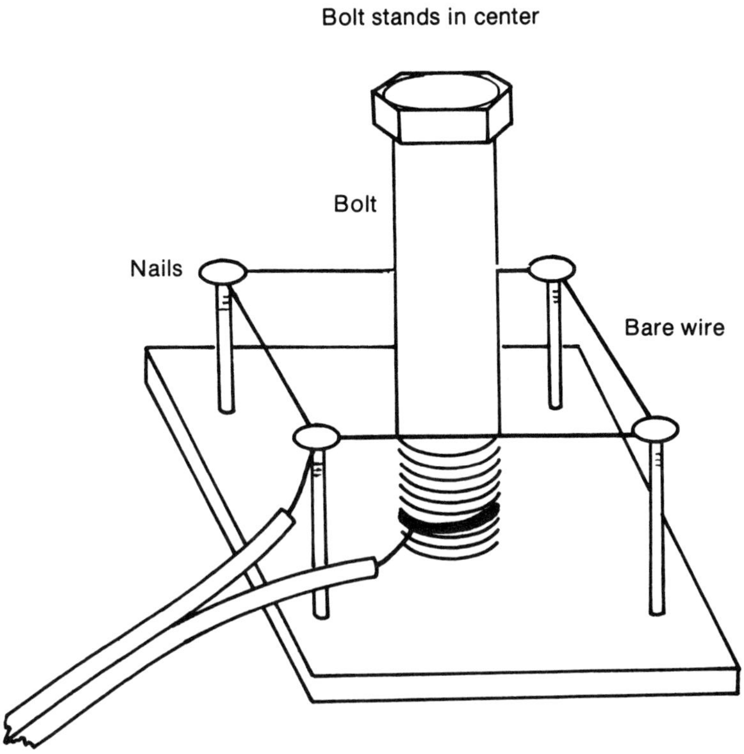

Fig. 68. The wire is raised up off the platform so that the bolt head will not prevent contact.

Simple Noise Maker

One of the easiest and fastest to make noise makers is a stack of breakable objects that will fall over and break on triggering. Anything made out of glass, such as drink glasses, bottles, etc., or ceramic, such as plates, mugs and so forth, will make noise when it falls and breaks. The farther it falls, the more noise it will make. It must fall against a hard surface. Fig. 69

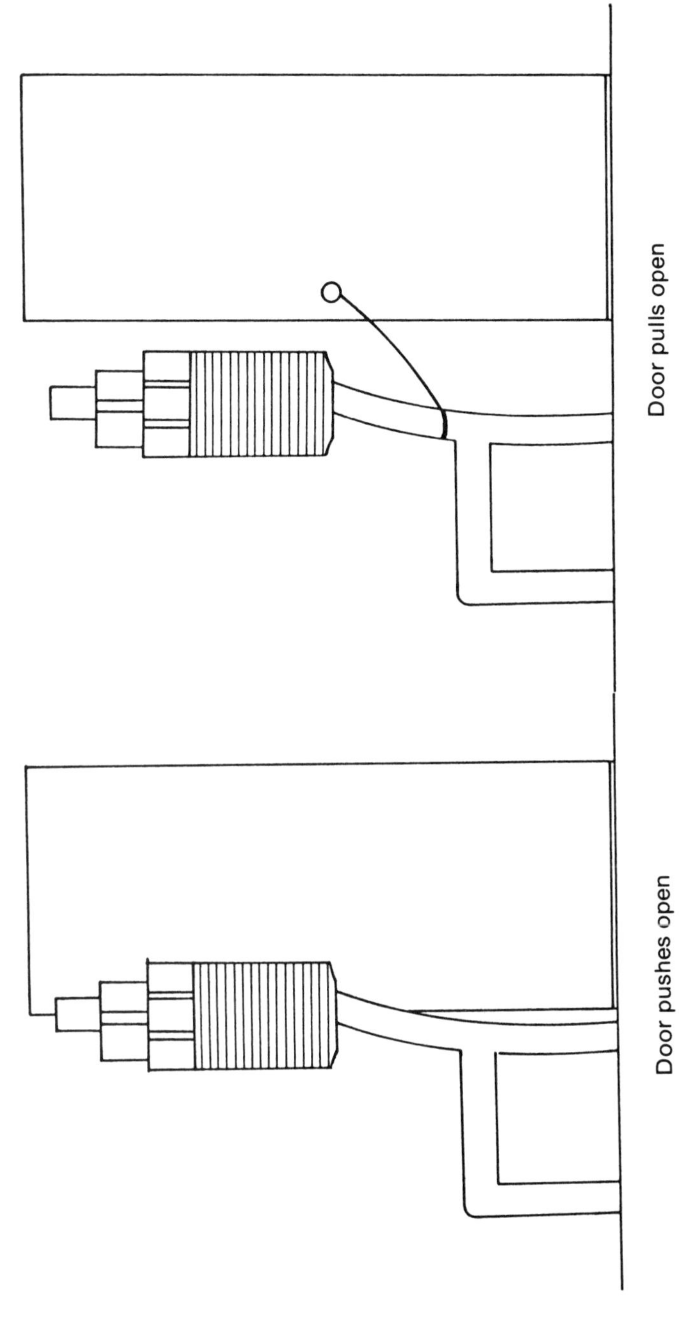

Fig. 69. Simple noise maker. Dishes are stacked and set to fall and break against a hard floor.

Bibliography

Army, Dept. of. **Boobytraps (FM-5-31)**. Washington: HQ, U.S. Army, 1965

Army, Dept. of. **Camouflage (FM 5-20)**. Washington: HQ, U.S. Army, 1968

Army, Dept. of. **Field Fortifications (FM 5-15)**. Washington: HQ, U.S. Army, 1972

Army, Dept. of. **An Infantryman's Guide to Urban Combat (FM 90-10-1)**. Washington: HQ, U.S. Army, 1982

-----**Guide to Viet Cong Boobytraps and Explosive Devices**. Cornville, AZ: Desert Publications, 1978

Frankford Arsenal. **Improvised Munitions Handbook**. Rosemead, CA: J. Flores Publications, 1969

Bateman, James A. **Animal Traps and Trapping**. Harrisburg, PA: Stackpole Books, 1971

Graves, Richard. **Bushcraft**. New York, NY: Schocken Books, 1972

Mason, Bernard S. **Woodcraft**. New York, NY: A.S. Barnes and Co., 1939

Cowper, H.S. **Art of Attack**. Ulverston: W. Holmes, LTD, 1906

-----**Home Security**. Time-Life Books, 1979

National Safety Council, Inc. **Accident Prevention Manual for Industrial Operations**. U.S., 1951

Miner, Douglas F., and Seastone, John B., editors. **Handbook of Engineering Materials**. New York: John Wiley and Sons, Inc., 1955

Hirth, Walter P. **Rigging**. Scranton, PA: Intertext, Inc., 1980

Day, Richard, et al. **Reader's Digest Complete Do-it Yourself Manual**. U.S. Reader's Digest Assoc., Inc., 1973

Gewirtz, Herman. **Essentials of Physics**. Woodbury, NY: Barron's Educational Series, Inc., 1971

Ilin, M. **Turning Night Into Day-The Story of Lighting.** Philadelphia, PA: J.B. Lippincott Co., 1936

Notes

Notes